Glencoe McGraw-Hill

Study Notebook

Precalculus

Mc Graw Hill Glencoe

The McGraw·Hill Companies

Glencoe

Send all inquiries to:
Glencoe/McGraw-Hill
8787 Orion Place
Columbus, OH 43240

ISBN: 978-0-07-893814-6
MHID: 0-07-893814-7

Printed in the United States of America

17 18 QTN 18 17 16

Contents

Note-Taking Tips

Your notes are a reminder of what you learned in class. Taking good notes can help you succeed in mathematics. The following tips will help you take better classroom notes.

- Before class, ask what your teacher will be discussing in class. Review mentally what you already know about the concept.
- Be an active listener. Focus on what your teacher is saying. Listen for important concepts. Pay attention to words, examples, and/or diagrams your teacher emphasizes.
- Write your notes as clear and concise as possible. The following symbols and abbreviations may be helpful in your note-taking.

Word or Phrase	Symbol or Abbreviation	Word or Phrase	Symbol or Abbreviation
for example	e.g.	not equal	$\neq$
such as	i.e.	approximately	$\approx$
with	w/	therefore	$\therefore$
without	w/o	versus	vs
and	+	angle	$\angle$

- Use a symbol such as a star (★) or an asterisk (∗) to emphasis important concepts. Place a question mark (?) next to anything that you do not understand.
- Ask questions and participate in class discussion.
- Draw and label pictures or diagrams to help clarify a concept.
- When working out an example, write what you are doing to solve the problem next to each step. Be sure to use your own words.
- Review your notes as soon as possible after class. During this time, organize and summarize new concepts and clarify misunderstandings.

Note-Taking Don'ts

- Don't write every word. Concentrate on the main ideas and concepts.
- Don't use someone else's notes as they may not make sense.
- Don't doodle. It distracts you from listening actively.
- Don't lose focus or you will become lost in your note-taking.

NAME _______________ DATE ________ PERIOD ________

CHAPTER 1

Functions from a Calculus Perspective

Before You Read

Before you read the chapter, respond to the following statements.

1. Write an **A** if you agree with the statement.
2. Write a **D** if you disagree with the statement.

Before You Read	Functions from a Calculus Perspective
	• For a given function *f*, a value in the domain is represented by the dependent variable *x* and a value in the range of *f* is represented by the independent variable *y*.
	• The graph of a continuous function has no breaks, holes, or gaps.
	• A function *f* is increasing on an interval if and only if for any two points in the interval, a positive change in *x* results in a negative change in *f*(*x*).
	• A translation is a rigid transformation that has the effect of shifting the graph of a function up, down, left, or right.
	• The inverse relation of a function is not necessarily a function.

Note-Taking Tips

- **In addition to writing important definitions in your notes, be sure to include your own examples of the concepts presented.**

  For example, when studying functions, be sure to include relations that are functions and relations that are not functions in your notes.

- **When you take notes, make sure that someone who did not understand the topic would understand after reading what you have written.**

NAME ______________________ DATE ____________ PERIOD ____________

CHAPTER 1 Functions from a Calculus Perspective

Key Points

Scan the pages in the chapter. Write at least one specific fact concerning each lesson. For example, in the lesson on functions, one fact might be that the rational and irrational number subsets form the real number set. After completing the chapter, you can use this table to review for your chapter test.

Lesson	Fact
1-1 Functions	
1-2 Analyzing Graphs of Functions and Relations	
1-3 Continuity, End Behavior, and Limits	
1-4 Extrema and Average Rates of Change	
1-5 Parent Functions and Transformations	
1-6 Function Operations and Composition of Functions	
1-7 Inverse Relations and Functions	

NAME ______________________ DATE __________ PERIOD __________

1-1 Functions

What You'll Learn

Scan Lesson 1-1. Write two things that you already know about functions.

1. ______________________________

____

2. ______________________________

____

Lesson 1-1

Active Vocabulary

New Vocabulary Write the definition next to each term.

domain ▶ ______________________________

____

dependent variable ▶ ______________________________

____

function ▶ ______________________________

____

independent variable ▶ ______________________________

____

range ▶ ______________________________

____

relation ▶ ______________________________

____

Main Idea	Details

Describe Subsets of Real Numbers

pp. 4–5

Complete the table. Write each set of numbers in set-builder and interval notation, if possible.

Set	Set-Builder Notation	Interval Notation
$\{-2, -1, 0, 1, \ldots\}$		
$x \leq 4$		
$-5 \leq x < 22$		
$x < -4$ or $x > 6$		
all multiples of 7		

Identify Functions

pp. 5–8

Determine whether each relation represents *y* as a function of *x*. Write *yes* or *no*.

1. $\{(2, -4), (-3, 7), (23, -5), (-3, 10)\}$ — 1. ________

2. The input value x is a Social Security number and the output value y is the owner's first name. — 2. ________

3. 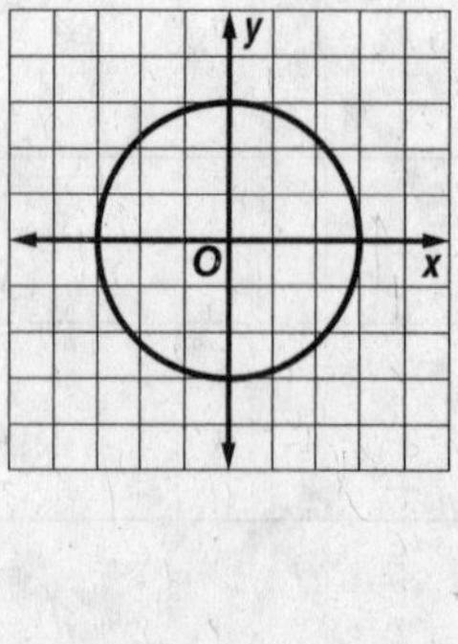

4.

x	y
−2	1
−1	0
0	0
1	1
2	3

3. ________

4. ________

5. $y = x^2 + 3x - 4$ — 5. ________

1-2 Analyzing Graphs of Functions and Relations

What You'll Learn

Scan the Examples for Lesson 1-2. Predict two things that you think you will learn about functions and their graphs.

1. ______________________________

____

____

2. ______________________________

____

____

Active Vocabulary

New Vocabulary Match the term with its definition by drawing a line to connect the two.

even function	the x-intercept(s) of the graph of a function
line of symmetry	functions that are symmetric with respect to the origin
odd function	graphs that have this property can be rotated 180° with respect to a point and appear unchanged
point symmetry	the solution(s) of a given equation
roots	functions that are symmetric with respect to the y-axis
zeros	graphs that have this property can be folded along a line so that the two halves of the graph match exactly

Main Idea	Details

Analyzing Function Graphs
pp. 13–16

Use the graph of each function to find its y-intercept and zeros.

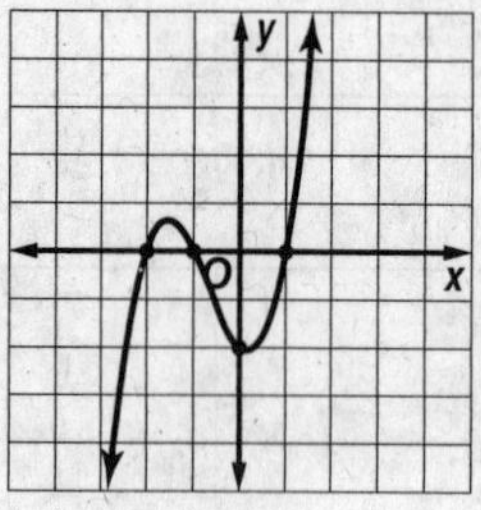

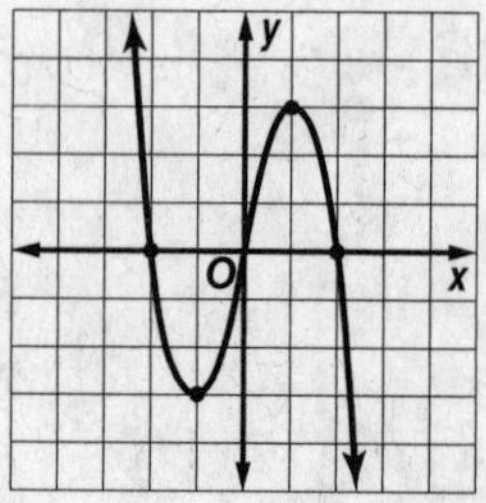

y-intercept: ____________ y-intercept: ____________

zeros: ____________ zeros: ____________

Symmetry of Graphs
pp. 16–18

Identify which function is *even*, which is *odd*, and which is *neither*.

1. $f(x) = x^3 - x$ 1. ____________
2. $g(x) = 2x^4 + x - 1$ 2. ____________
3. $h(x) = -3x^2 + 1$ 3. ____________

Helping You Remember

Think about the different types of line symmetry: about the x-axis, about the y-axis, and about the origin. Provide examples of graphs illustrating each of the symmetries.

x-axis symmetry

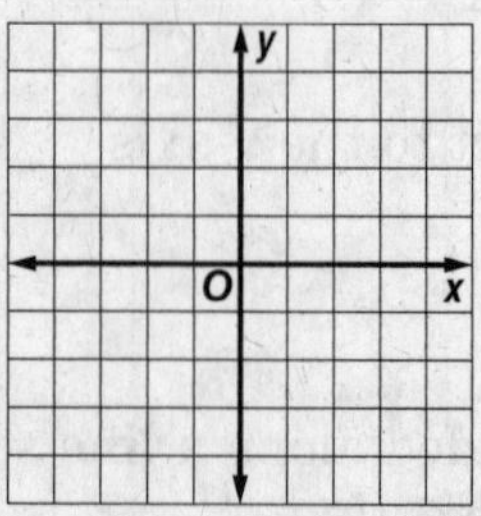

y-axis symmetry

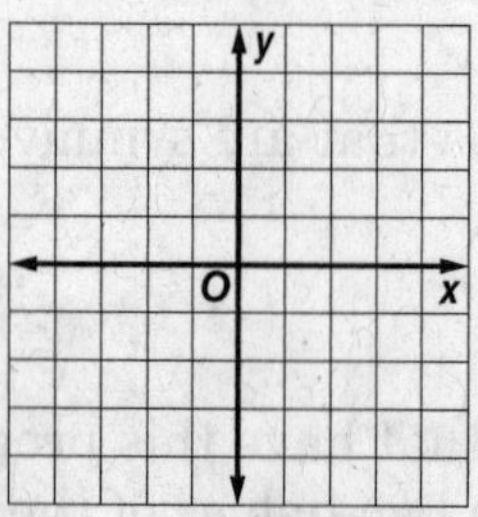

origin symmetry

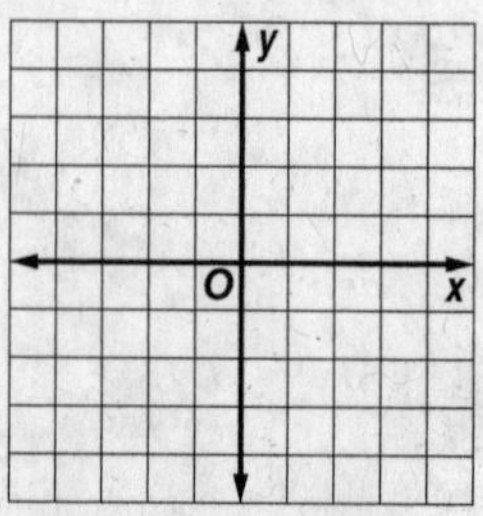

NAME ______________________ DATE ____________ PERIOD ____________

1-3 Continuity, End Behavior, and Limits

What You'll Learn

Scan the text in Lesson 1-3. Write two facts that you learned about continuity.

1. ____

____

____

2. ____

____

____

Active Vocabulary

New Vocabulary Write the correct term next to each definition.

______________ ▶ a function that has no breaks, holes, or gaps in its graph

______________ ▶ a concept describing how a function behaves at either end of its graph

______________ ▶ a function is said to have this form of a discontinuity at $x = c$ if the absolute value of the function increases or decreases indefinitely as the x-values approach c from the left and the right

______________ ▶ the concept of approaching a value without necessarily ever reaching it

______________ ▶ a function is said to have this form of discontinuity at $x = c$ if the function is continuous everywhere except for a hole at $x = c$

______________ ▶ a function that is not continuous

______________ ▶ a function is said to have this form of discontinuity at $x = c$ if the limits of the function as x approaches c from the left and the right exist but have two distinct values

Lesson 1-3

Main Idea | **Details**

Continuity
pp. 24–26

Complete the table by providing your own verbal description of each type of discontinuity. Then provide an example to illustrate your verbal description.

Discontinuity	Verbal Description	Example
Infinite		
Jump Discontinuity		
Removable or Point		

End Behavior
pp. 28–29

Use the graph of $f(x) = x^3 - 2x^2 - 5x + 6$ to describe its end behavior.

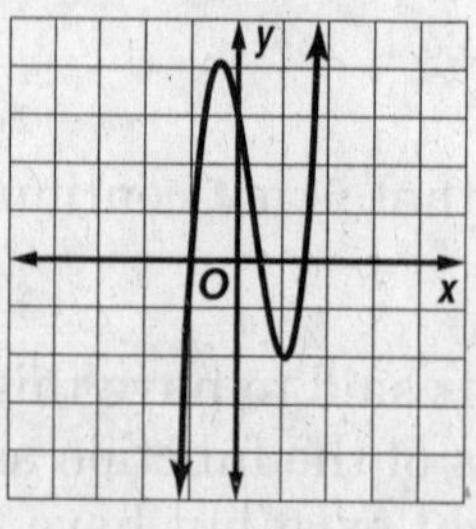

____

____

____

NAME ______________________ DATE ____________ PERIOD ____________

1-4 Extrema and Average Rates of Change

What You'll Learn

Scan the text under the *Now* heading. List two things that you will learn in this lesson.

1. ____

2. ____

Active Vocabulary

New Vocabulary Label the diagram with the terms listed at the left.

constant

decreasing

increasing

maximum

minimum

secant line

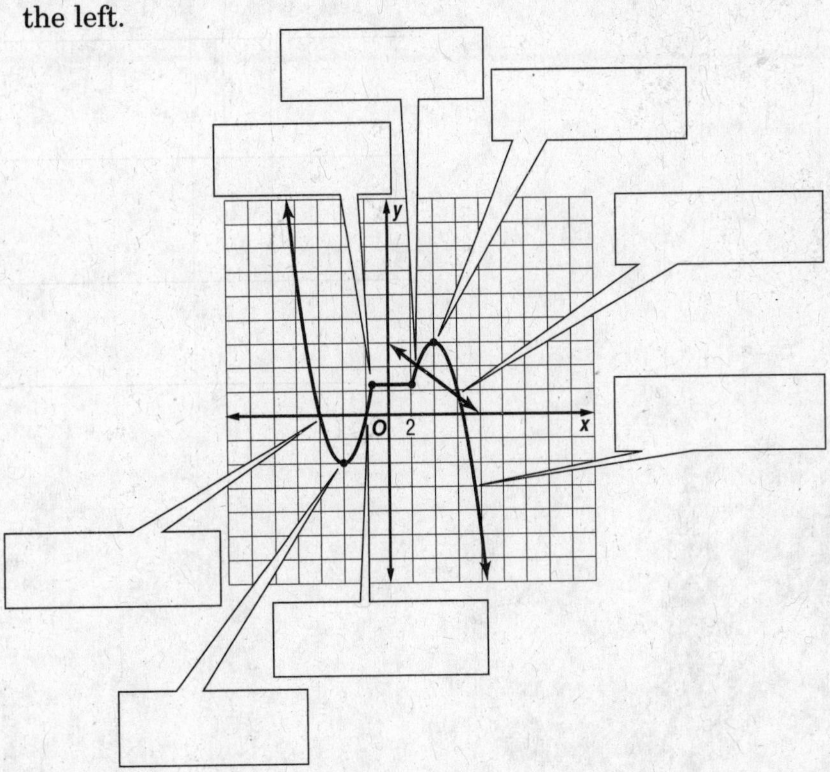

Lesson 1-4

Main Idea	Details

Increasing and Decreasing Behavior
pp. 34–38

Draw the graph of a function modeling the indicated behavior throughout its domain.

Increasing Function	Decreasing Function	Constant Function

Average Rate of Change
pp. 38–39

Find the average rate of change of $f(x) = x^4 - 3x^2 + 6x$ on the interval $[-1, 2]$.

______________________ Slope formula

______________________ Substitute −1 for x_1 and 2 for x_2.

______________________ Evaluate $f(3)$ and $f(-2)$.

______________________ Simplify.

1-5 Parent Functions and Transformations

What You'll Learn

Scan Lesson 1-5. Predict two things that you expect to learn based on the headings and Key Concept boxes.

1. ______________________________

____

____

2. ______________________________

____

____

Active Vocabulary

New Vocabulary Fill in each blank with the correct term.

absolute value function — A(n) ______________________ is a rigid transformation that has the affect of shifting the graph of a function up, down, left, or right.

identity function — A(n) ______________________ is the simplest of the functions in a family.

parent function — A(n) ______________________ is a rigid transformation which produces a mirror image of the graph of a function with respect to a specific line.

reflection — The ______________________ $f(x) = x$ passes through all points with coordinates (a, a).

transformations — A(n) ______________________ of a parent function affects the appearance of the parent graph.

translation — The ______________________, denoted as $f(x) = |x|$, is a V-shaped function.

Lesson 1-5

Lesson 1-5 *(continued)*

Main Idea	Details

Parent Functions
pp. 45–46

Each graph is the parent function for a family of functions. Identify the parent function.

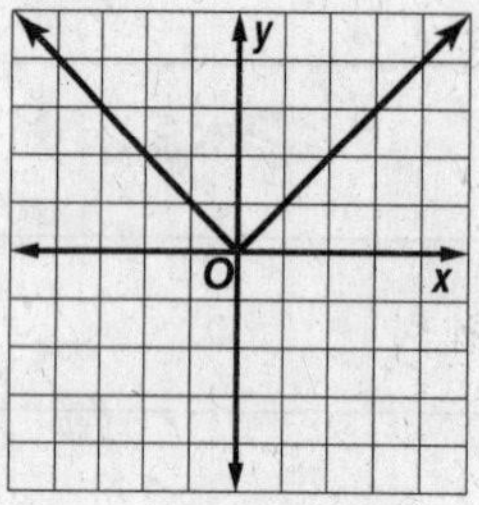

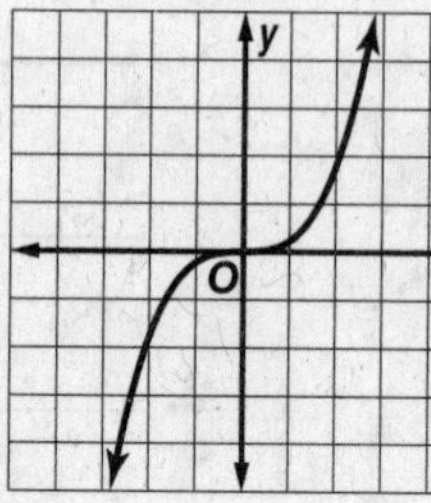

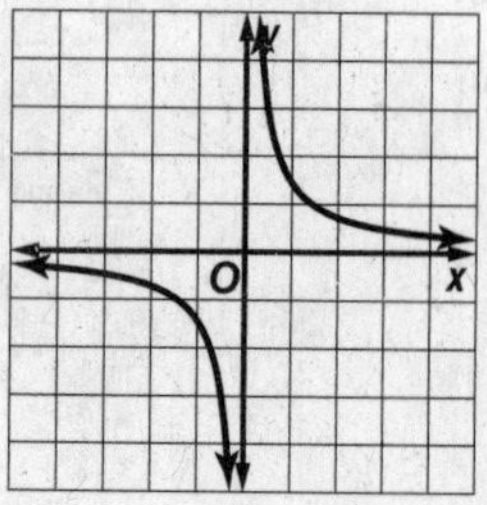

______________ ______________ ______________

Transformations
pp. 46–51

Define the three different transformations introduced in this lesson.

Translation ______________________________

____

____

Reflection ______________________________

____

____

Dilation ______________________________

____

____

____

NAME ______________________ DATE ____________ PERIOD ____________

1-6 Function Operations and Composition of Functions

What You'll Learn

Scan Lesson 1-6. List two headings that you would use to make an outline of this lesson.

1. ______________________

____

2. ______________________

____

Active Vocabulary

Review Vocabulary Define *function* in your own words. (Lesson 1-1)

function ▶ ______________________

____

____

Define *relation* in your own words. (Lesson 1-1)

relation ▶ ______________________

____

Define *roots* in your own words. (Lesson 1-2)

roots ▶ ______________________

____

New Vocabulary Write the definition next to the term.

composition ▶ ______________________

____

Main Idea	Details

Operations with Functions
pp. 57–58

Given $f(x) = 2x + 1$ and $g(x) = 4x^2 - 1$, find each function and its domain.

1. $(f + g)(x) =$ ____________________

   Domain: ____________________

2. $(f - g)(x) =$ ____________________

   Domain: ____________________

3. $(f \cdot g)(x) =$ ____________________

   Domain: ____________________

4. $\left(\frac{f}{g}\right)(x) =$ ____________________

   Domain: ____________________

Composition of Functions
pp. 58–60

Given $f(x) = x + 1$, $g(x) = x^2 - 2$, and $h(x) = x^2 + 3x - 4$, find each composition.

1. $[f \circ g](x) =$ ____________________

2. $g[f(x)] =$ ____________________

3. $[h \circ f](x) =$ ____________________

Helping You Remember

In the composition $f \circ g$, which is read as f composition g or f of g, the function g is applied first then f. Think of a mnemonic device for remembering how to find the composition of two functions f and g so that you are not confused when you see $f \circ g$ or $g \circ f$.

____

____

____

NAME ______________________ DATE __________ PERIOD __________

1-7 Inverse Relations and Functions

What You'll Learn

Scan the Examples for Lesson 1-7. Predict two things that you think you will learn about inverse relations.

1. ______________________

____

2. ______________________

____

Active Vocabulary

Review Vocabulary Define *domain* in your own words. (Lesson 1-1)

domain ▶ ______________________

____

Define *range* in your own words. (Lesson 1-1)

range ▶ ______________________

____

New Vocabulary Fill in each blank with the correct term or phrase.

inverse relations — If a function passes the horizontal line test, then it is said to be ______________, because no x-value is matched with more than one y-value and no y-value is matched with more than one x-value.

inverse function — Two relations are ______________ if and only if one relation contains the element (b, a) whenever the other relation contains the element (a, b).

one-to-one — If the inverse relation of a function f is also a function, then it is called the ______________ of f.

Lesson 1-7

Main Idea	Details
Inverse Functions pp. 65–66	**Graph each function using a graphing calculator, and apply the horizontal line test to determine whether its inverse function exists. Write *yes* or *no*.**

1. $f(x) = x^3 + 1$ ____________
2. $g(x) = \frac{3}{2 - x}$ ____________
3. $h(x) = -2|x - 4| + 1$ ____________
4. $g(x) = \frac{x + 2}{x - 4}$ ____________
5. $f(x) = x^3 + x^2 - 3x$ ____________

Find Inverse Functions
pp. 66–69

Find the inverse of $f(x) = \frac{x + 3}{x - 2}$.

______________________ Original function

______________________ Replace $f(x)$ with y.

______________________ Exchange x and y.

______________________ Solve for y.

______________________ Replace y with $f^{-1}(x)$.

Helping You Remember

In Lesson 1-6, you learned how to find the composition of two functions. Explain what role the composition of functions plays in determining whether two functions are inverses of one another.

____

____

CHAPTER 1

Functions from a Calculus Perspective

Tie It Together

Use the graph of $f(x)$ to complete the graphic organizer.

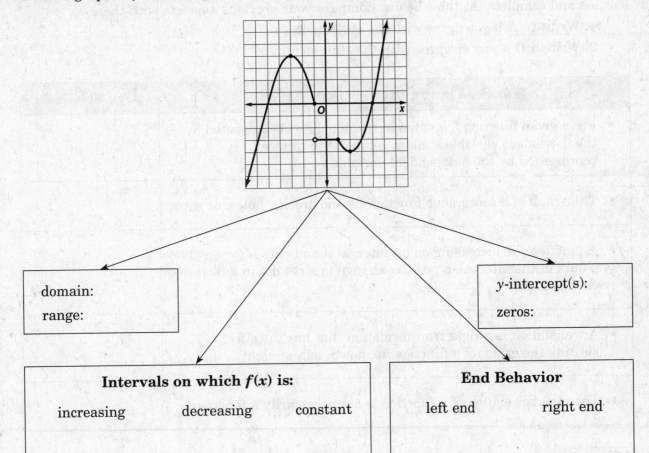

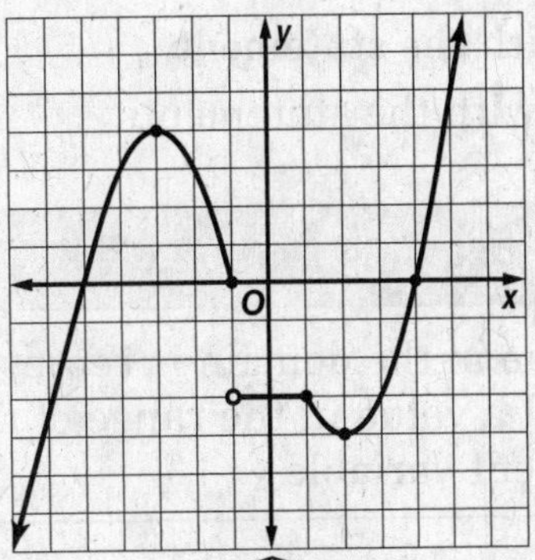

domain:

range:

y-intercept(s):

zeros:

Intervals on which $f(x)$ is:

increasing | decreasing | constant

End Behavior

left end | right end

Determine whether $f(x)$ has each of the following characteristics. Write *yes* or *no*. If yes, state the x-values where they occur.

Point(s) of Discontinuity

infinite | jump | point

Extrema

relative maximum | absolute maximum

relative minimum | absolute minimum

NAME ______________________ DATE __________ PERIOD __________

CHAPTER 1 Functions from a Calculus Perspective

Before the Test

Now that you have read and worked through the chapter, think about what you have learned and complete the table below. Compare your previous answers with these.

1. Write an **A** if you agree with the statement.
2. Write a **D** if you disagree with the statement.

Functions from a Calculus Perspective	After You Read
• For a given function *f*, a value in the domain is represented by the dependent variable *x* and a value in the range of *f* is represented by the independent variable *y*.	
• The graph of a continuous function has no breaks, holes, or gaps.	
• A function *f* is increasing on an interval if and only if for any two points in the interval, a positive change in *x* results in a negative change in $f(x)$.	
• A translation is a rigid transformation that has the effect of shifting the graph of a function up, down, left, or right.	
• The inverse relation of a function is not necessarily a function.	

Math Online Visit *glencoe.com* to access your textbook, more examples, self-check quizzes, personal tutors, and practice tests to help you study for concepts in Chapter 1.

Are You Ready for the Chapter Test?

Use this checklist to help you study.

☐ I completed the Chapter 1 Study Guide and Review in the textbook.

☐ I took the Chapter 1 Practice Test in the textbook.

☐ I used the online resources for additional review options.

☐ I reviewed my homework assignments and made corrections to incorrect answers.

☐ I reviewed all vocabulary terms from the chapter.

Study Tip

- When studying for tests, create and use graphic organizers to show relationships between concepts.

NAME ______________________ DATE ____________ PERIOD ____________

CHAPTER 2

Power, Polynomial, and Rational Functions

Before You Read

Before you read the chapter, respond to these statements.

1. Write an **A** if you agree with the statement.
2. Write a **D** if you disagree with the statement.

Before You Read	Power, Polynomial, and Rational Functions
	• When solving a radical equation, you should check for extraneous solutions.
	• The graph of a polynomial function may contain breaks, holes, gaps, or sharp corners.
	• Synthetic division is a shortcut for dividing a polynomial by a linear factor of the form $x - c$.
	• A polynomial function of degree n, where $n > 0$ has at least one zero in the real number system.
	• If the degree of the numerator of a rational function is the same as the degree of the denominator, then the graph will have no horizontal asymptote.

Note-Taking Tips

- **In addition to writing definitions in your notes, be sure to include examples and graphs to display the concepts presented.**
  For example, when studying rational functions be sure to include notes on how to find the vertical asymptote(s), horizontal asymptote, intercepts, and holes.
- **Take notes that are specific enough that you will understand what they mean when you read them a few days later.**

NAME ______________________ DATE ____________ PERIOD ____________

CHAPTER 2

Power, Polynomial, and Rational Functions

Key Points

Scan the pages in the chapter. Write at least one specific fact concerning each lesson. For example, in the lesson on power and radical functions, one fact might be to describe the shape of the graph of an even-degree and an odd-degree radical function. After completing the chapter, you can use this table to review for your chapter test.

Lesson	Fact
2-1 Power and Radical Functions	
2-2 Polynomial Functions	
2-3 The Remainder and Factor Theorems	
2-4 Zeros of Polynomial Functions	
2-5 Rational Functions	
2-6 Nonlinear Inequalities	

NAME ______________________ DATE ____________ PERIOD ____________

2-1 Power and Radical Functions

What You'll Learn

Scan Lesson 2-1. Predict two things that you expect to learn based on the headings and Key Concept box.

1. ______________________________

____

____

2. ______________________________

____

____

Active Vocabulary

New Vocabulary Fill in the blank with the correct term.

extraneous solution A ______________ is any function of the form $f(x) = ax^n$ where a and n are nonzero constant real numbers.

monomial function A ______________ is a power function in which n is a positive integer.

power function A ______________ is a function that has at least one radical expression containing the independent variable.

radical function Solutions that do not satisfy the original equation are called ______________.

Review Vocabulary Define *function* in your own words. (Lesson 1-1)

function ______________________________

____

____

Lesson 2-1

Main Idea	Details

Power Functions
pp. 86–89

Identify the power functions in this list by placing each function in the correct box.

$f(x) = 2x^3$ $\qquad$ $f(x) = 3^x$

$f(x) = \frac{1}{x}$ $\qquad$ $f(x) = -5x^{\frac{2}{3}}$

$f(x) = 4x - 1$ $\qquad$ $f(x) = \frac{1}{5 - x}$

Power Functions	Not Power Functions

Radical Functions
pp. 89–91

Find the domain, range, and intercepts of $f(x) = 2\sqrt[3]{8 - 3x}$.

domain: ____________ x-intercept: ____________

range: ____________ y-intercept: ____________

Helping You Remember

How can you remember the domain of $f(x) = x^{\frac{p}{n}}$ where n is even or n is odd? Sketch a graph of each.

____

____

***n* is even.**

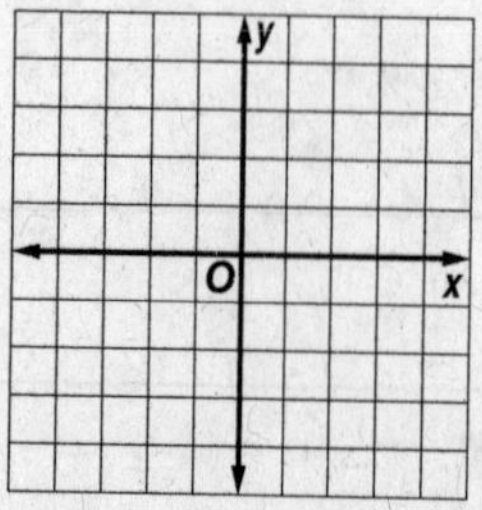

***n* is odd.**

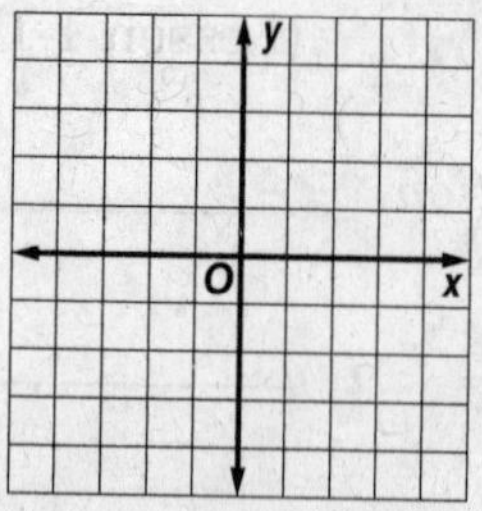

2-2 Polynomial Functions

What You'll Learn

Scan the Examples for Lesson 2-2. Predict two things that you think you will learn about polynomial functions.

1. ______________________________

____

____

2. ______________________________

____

____

Active Vocabulary

New Vocabulary Match the term with its definition by drawing a line to connect the two.

polynomial function	the coefficient of the variable with the greatest exponent
leading coefficient	the zero of a factor $(x - c)$ that occurs more than once in the completely factored form of $f(x)$
turning point	the number of times a zero is repeated
repeated zero	functions formed from the sums and differences of monomial functions and constants
multiplicity	points that indicate where the graph of the function changes from increasing to decreasing, and vice versa

NAME ______________________________ DATE ____________ PERIOD ____________

Main Idea	Details

Graph Polynomial Functions
pp. 97–102

Complete the chart for the Leading Term Test for Polynomial End Behavior.

n is odd, a_n is positive. $\lim_{x \to -\infty} f(x) =$ ________ $\lim_{x \to \infty} f(x) =$ ________	n is odd, a_n is negative. $\lim_{x \to -\infty} f(x) =$ ________ $\lim_{x \to \infty} f(x) =$ ________
n is even, a_n is positive. $\lim_{x \to -\infty} f(x) =$ ________ $\lim_{x \to \infty} f(x) =$ ________	n is even, a_n is negative. $\lim_{x \to -\infty} f(x) =$ ________ $\lim_{x \to \infty} f(x) =$ ________

Model Data
p. 103

Write a polynomial function to model the number of DVD players sold during February in the United States as a function of the number of years *t* since 2000 as shown in the chart.

Years Since 2000	4	5	6	7
Number of DVD Players	919,295	590,128	770,132	1,117,899

____

Helping You Remember

Look up the words *quartet*, *quarter*, and *quarto*. How can the meaning of these words help you remember the meaning of *quartic function*?

____

____

NAME ______________________ DATE __________ PERIOD __________

2-3 The Remainder and Factor Theorems

What You'll Learn

Scan the lesson. Write two things that you already know about solving quadratic equations.

1. ______________________

2. ______________________

Active Vocabulary

Review Vocabulary Define *domain* in your own words. (Lesson 1-1)

domain ▶ ______________________

Define *range* in your own words. (Lesson 1-1)

range ▶ ______________________

New Vocabulary Write the correct term next to each definition.

______________ ▶ a shortcut for dividing a polynomial by a linear factor of the form $(x - c)$

______________ ▶ the process of using synthetic division to evaluate a function

______________ ▶ the quotient when a polynomial is divided by one of its binomial factors $(x - c)$

Lesson 2-3

Main Idea	Details
Divide Polynomials pp. 109–112	**Fill in each box with the division indicated.**

Long Division	Synthetic Division
$x + 3\overline{)3x^3 + 6x^2 + x - 2}$	$\underline{3\rfloor}$ 3 6 1 −2

The Remainder and Factor Theorems
pp. 112–114

Determine if each expression is a factor of $f(x) = x^3 - 2x^2 - 13x - 10$. Write *yes* or *no*.

1. $x - 2$ __________
2. $x + 1$ __________
3. $x + 3$ __________
4. $x - 5$ __________
5. $x + 2$ __________
6. $x - 1$ __________

Helping You Remember

Suppose the remainder when $f(x)$ is divided by $(x - c)$ is 0. Explain what this tells you about factors, zeros of the function, and x-intercepts of the graph.

____

____

____

NAME ______________________ DATE ____________ PERIOD ____________

2-4 Zeros of Polynomial Functions

What You'll Learn

Scan the text in Lesson 2-4. Write two facts that you learned about zeros of polynomial functions.

1. ______________________

2. ______________________

Active Vocabulary

New Vocabulary Write the definition next to each term.

Descartes' Rule of Signs ▶ ______________________

Fundamental Theorem of Algebra ▶ ______________________

Rational Zero Theorem ▶ ______________________

Lesson 2-4

Main Idea	Details

Real Zeros
pp. 119–123

Use the Rational Zero Theorem, the Upper and Lower Bound Tests, or Descartes' Rule of Signs to find the zeros of $f(x) = x^3 - 7x - 6$.

Find all possible rational zeros.	
Determine an interval within which all of the real zeros are located.	
Determine the number of positive and negative real zeros.	
Find the real zeros.	

Complex Zeros
pp. 123–126

Find the minimum degree of the polynomial function with real coefficients that have the given zeros.

1. $2, -7$ ________

2. $6, -6, \frac{1}{3}$ ________

3. $3 + \sqrt{7}, 2 - 8i$ ________

NAME ______________________ DATE __________ PERIOD __________

2-5 Rational Functions

What You'll Learn

Scan the text under the *Now heading*. List two things that you will learn in the lesson.

1. ____

2. ____

Active Vocabulary

New Vocabulary Label the diagram with the correct terms.

hole

horizontal asymptote

rational function

vertical asymptote

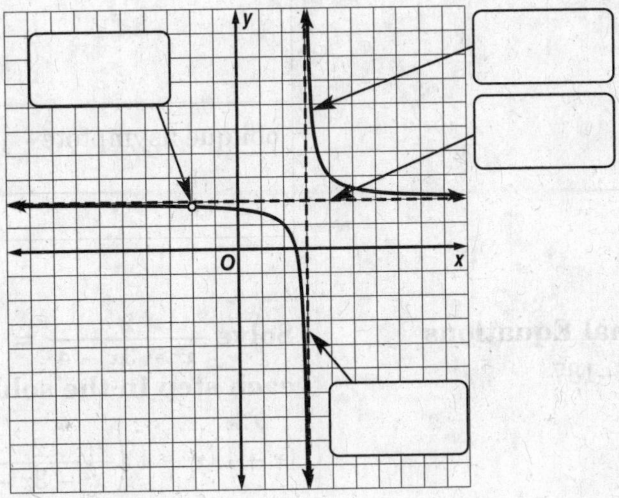

Vocabulary Link A synonym for *extraneous* is *irrelevant*. Explain how this relates to extraneous solutions.

extraneous ▶ ____

Main Idea | **Details**

Rational Functions
pp. 130–136

Explain how to find each characteristic for a rational function.

Characteristic	Method
hole(s)	
vertical asymptote(s)	
horizontal asymptote	
oblique asymptote	

Rational Equations
pp. 136–137

Solve $\frac{2}{x^2 - 3x - 4} = \frac{x}{x - 4} + \frac{2}{x + 1}$. Write the reason for each step in the solution.

$(x - 4)(x + 1)\left(\frac{2}{x^2 - 3x - 4}\right) = (x - 4)(x + 1)\left(\frac{x}{x - 4} + \frac{2}{x + 1}\right)$ ______________________

$2 = x(x + 1) + 2(x - 4)$ ______________________

$2 = x^2 + 3x - 8$ ______________________

$0 = x^2 + 3x - 10$ ______________________

$0 = (x + 5)(x - 2)$ ______________________

$x = -5$ or $x = 2$ ______________________

NAME ______________________________ DATE ____________ PERIOD ____________

2-6 Nonlinear Inequalities

Lesson 2-6

What You'll Learn

Scan Lesson 2-6. List two headings that you would use to make an outline of this lesson.

1. ______________________________

____

2. ______________________________

____

Active Vocabulary

New Vocabulary Fill in each blank with the correct term.

polynomial inequality

rational inequality

sign chart

The numbers used to make a sign chart are called ______________.

A ______________ has the general form $f(x) \leq 0$, $f(x) < 0$, $f(x) \neq 0$, $f(x) > 0$, or $f(x) \geq 0$.

A ______________ is an inequality formed using a rational function.

Review Vocabulary Define *polynomial function* in your own words. (Lesson 2-2)

polynomial function ▶ ______________________________

____

____

zeros ▶ Define *zeros* in your own words. (Lesson 1-2)

____

____

Main Idea	Details

Polynomial Inequalities
pp. 141–143

Complete the chart to solve $x^3 + 3x^2 - 6x - 8 > 0$.

Let $f(x) = x^3 + 3x^2 - 6x - 8$. Factor $f(x)$ and find all of the real zeros.	
Determine the end behavior of $f(x)$.	
Complete the sign chart.	
Write the solutions.	

Rational Inequalities
pp. 143–144

Solve $\frac{4}{x+5} < \frac{1}{2x+3}$. Complete the steps shown.

______________________ Subtract $\frac{1}{(2x+3)}$ from each side.

______________________ Use the LCD to rewrite each fraction. Then add.

______________________ Find the zeros and undefined points.

______________________ Complete the sign chart using these numbers.

______________________ Solution

NAME ______________________ DATE ____________ PERIOD ____________

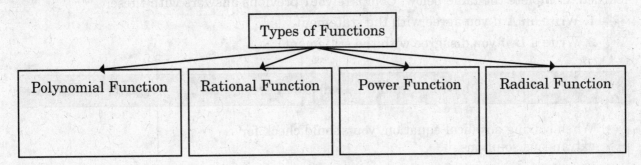

CHAPTER 2 Power, Polynomial, and Rational Functions

Tie It Together

Complete the graphic organizer by providing examples of each type of function.

Types of Functions

- Polynomial Function
- Rational Function
- Power Function
- Radical Function

Complete the organizer by providing an example of each inequality and its solution.

Nonlinear Inequalities

- Polynomial Inequality
- Rational Inequality

Explain how each theorem or rule is used.

Theorem / Rule	How it is used
Rational Zero Theorem	
Descartes' Rule of Signs	
Factor Theorem	
Remainder Theorem	

NAME ______________________ DATE ____________ PERIOD ____________

CHAPTER 2 Power, Polynomial, and Rational Functions

Before the Test

Now that you have read and worked through the chapter, think about what you have learned. Complete the table below. Compare your previous answers with these.

1. Write an **A** if you agree with the statement.

2. Write a **D** if you disagree with the statement.

Power, Polynomial, and Rational Functions	After You Read
• When solving a radical equation, you should check for extraneous solutions.	
• The graph of a polynomial function may contain breaks, holes, gaps, or sharp corners.	
• Synthetic division is a shortcut for dividing a polynomial by a linear factor of the form $x - c$.	
• A polynomial function of degree n, where $n > 0$ has at least one zero in the real number system.	
• If the degree of the numerator of a rational function is the same as the degree of the denominator, then the graph will have no horizontal asymptote.	

Math Online Visit *glencoe.com* to access your textbook, more examples, self-check quizzes, personal tutors, and practice tests to help you study for concepts in Chapter 2.

Are You Ready for the Chapter Test?

Use this checklist to help you study.

- ☐ I completed the Chapter 2 Study Guide and Review in the textbook.
- ☐ I took the Chapter 2 Practice Test in the textbook.
- ☐ I used the online resources for additional review options.
- ☐ I reviewed my homework assignments and made corrections to incorrect answers.
- ☐ I reviewed all vocabulary terms from the chapter.

Study Tip

- Review your homework assignments. Redo any problems that you missed when they were first assigned.

CHAPTER 3

Exponential and Logarithmic Functions

Before You Read

Before you read the chapter, respond to these statements.

1. Write an **A** if you agree with the statement.
2. Write a **D** if you disagree with the statement.

Before You Read	Exponential and Logarithmic Functions
	• The function $f(x) = x^3$ is an exponential function.
	• A natural logarithm is a logarithm with base e.
	• $\log_b x + \log_b y = \log_b (x + y)$
	• $36^{x+1} = 6^{(x+1)^2}$
	• Data exhibiting rapid growth or decay can be modeled using a logarithmic function.

Note-Taking Tips

- **Before each lesson, scan through the lesson. Write any questions that you think of in your notes.**
  As you study the lesson, record the answers to your questions.
- **Always write clear, concise notes so they can be easily read when studying for a quiz or test.**

NAME ______________________ DATE ____________ PERIOD ____________

CHAPTER 3 Exponential and Logarithmic Functions

Key Points

Scan the pages in the chapter. Write at least one specific fact concerning each lesson. For example, in the lesson on logarithmic functions, one fact might be that a logarithmic function is the inverse of an exponential function. After completing the chapter, you can use this table to review for your chapter test.

Lesson	Fact
3-1 Exponential Functions	
3-2 Logarithmic Functions	
3-3 Properties of Logarithms	
3-4 Exponential and Logarithmic Equations	
3-5 Modeling with Nonlinear Regression	

NAME ______________________ DATE ____________ PERIOD ____________

3-1 Exponential Functions

What You'll Learn

Scan Lesson 3-1. Predict two things that you expect to learn based on the headings and figures in the lesson.

1. ______________________________

____

2. ______________________________

____

Lesson 3-1

Active Vocabulary

New Vocabulary Write the definition next to each term.

algebraic function ▶ ______________________________

____

____

____

continuous compound interest ▶ ______________________________

____

exponential function ▶ ______________________________

____

natural base ▶ ______________________________

____

transcendental function ▶ ______________________________

____

Main Idea	Details

Exponential Functions
pp. 158–161

Complete the chart.

Exponential Function	Growth	Decay
domain		
range		
x-intercept		
y-intercept		
asymptote		
end behavior		

Exponential Growth and Decay
pp. 161–165

Complete the steps in the solution of the problem below.

Beth invests $500 at a 4% interest rate that is compounded quarterly. If she makes no other deposits or withdrawals, what will her balance be after 10 years?

______________ Compound Interest Formula

______________ Substitute.

______________ Simplify.

After 10 years, her balance will be __________.

NAME ______________________ DATE ____________ PERIOD ____________

3-2 Logarithmic Functions

What You'll Learn

Scan the Examples for Lesson 3-2. Predict two things that you think you will learn about logarithmic functions.

1. ____

____

2. ____

____

Active Vocabulary

New Vocabulary Label the diagram with the correct terms.

base

common logarithm

logarithm

logarithmic function with base b

natural logarithm

exponent

$f(x) = \log_b x$ } → [ ]

↓ [ ]

$\log_{10} x$ } → [ ]

$\ln x$ } → [ ]

$\log_2 8 = 3$ } → [ ]

↓ [ ]

Lesson 3-2

Main Idea	Details
Logarithmic Functions and Expressions pp. 172–174	**Evaluate each expression.**

1. $\log_4 16$ ________ **2.** $\ln e^5$ ________

3. $\log_2 \frac{1}{8}$ ________ **4.** $\log_8 8$ ________

5. $\log_6 \sqrt[3]{6}$ ________ **6.** $\ln \left(\frac{1}{e^4}\right)$ ________

Graphs of Logarithmic Functions
pp. 175–177

Use the graph of $f(x) = \log x$ to describe the transformation that results in each function. Then sketch the graphs of the functions.

1. $g(x) = \log (x - 2)$

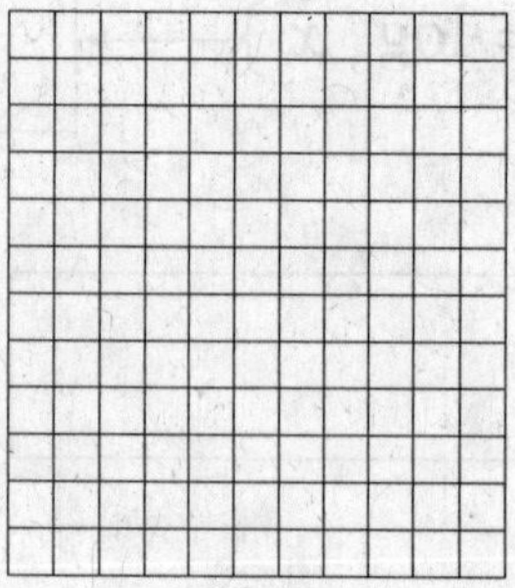

2. $h(x) = 2 \log x + 1$

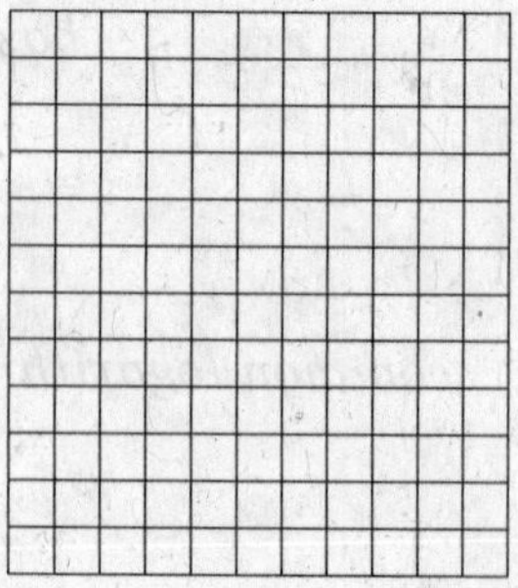

Helping You Remember

Write an example that can help you remember how to change an equation from logarithmic form to exponential form.

____

____

____

NAME ______________________ DATE ____________ PERIOD ____________

3-3 Properties of Logarithms

What You'll Learn

Scan the text in Lesson 3-3. Write two facts that you learned about properties of logarithms as you scanned the text.

1. ______________________

____

2. ______________________

____

____

Active Vocabulary

Review Vocabulary Use each term in a sentence that shows its mathematical meaning.

exponential function ▶ ______________________

____

natural logarithm ▶ ______________________

____

transcendental function ▶ ______________________

____

common logarithm ▶ ______________________

____

Lesson 3-3

Main Idea | **Details**

Properties of Logarithms
pp. 181–183

Fill in the table with an example of each property.

Property	Example
Product Property	
Quotient Property	
Power Property	

Change of Base Formula
pp. 183–184

Apply the change of base formula to evaluate each logarithm. Round answers to the nearest hundredth.

1. $\log_4 6$ ____________

2. $\log_3 96$ ____________

3. $\log_{\frac{1}{4}} 2$ ____________

4. $\log_{51} 26$ ____________

Helping You Remember

To help you remember that the log of a product is equal to the sum of the logs, relate this fact to a property of exponents. Give an example of each relationship.

____

____

____

NAME ________________ DATE ________ PERIOD ________

3-4 Exponential and Logarithmic Equations

What You'll Learn

Scan Lesson 3-4. List two that headings that you would use to make an outline of this lesson.

1. ______________________________

2. ______________________________

Active Vocabulary

Review Vocabulary List the three properties of exponents that you learned in Lesson 3-3. Provide an example of each property.

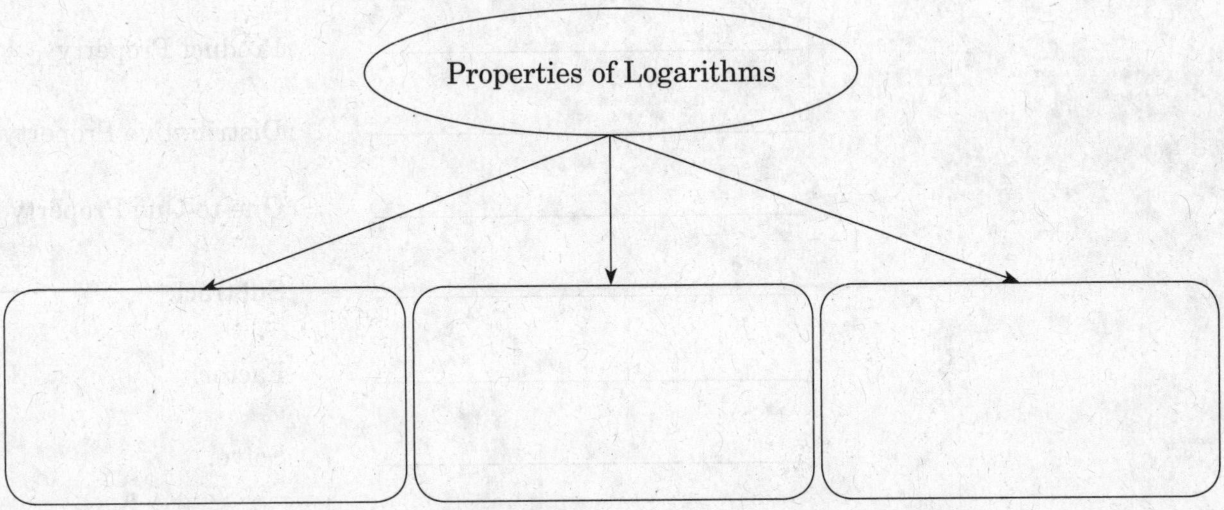

Vocabulary Link When you solved radical equations, you needed to watch for extraneous solutions that would cause the expression under the radical to be negative. What extraneous solutions might occur when solving a logarithmic equation?

____

____

____

____

Main Idea	Details

One-to-One Property of Exponential Functions
pp. 190–191

Solve each equation.

1. $4^x = 16^{x-2}$ ______________

2. $\left(\frac{1}{3}\right)^a = 27^4$ ______________

3. $\log_5 \sqrt[3]{25} = x$ ______________

4. $\ln\left(\frac{1}{x}\right) = \frac{1}{2}$ ______________

One-to-One Property of Logarithmic Functions
pp. 191–195

Complete the steps to solve $\log_2 10 = \log_2 x + \log_2 (x - 3)$.

______________ Original equation

______________ Product Property

______________ Distributive Property

______________ One-to-One Property

______________ Subtract.

______________ Factor.

______________ Solve.

______________ Check solutions.

Helping You Remember

Many students have trouble recognizing when they can solve an exponential equation using the One-to-One Property and when they must use logarithms. Explain how they could remember this.

____

____

____

3-5 Modeling with Nonlinear Regression

What You'll Learn

Scan the lesson. Write two things that you already know about modeling data.

1. ______________________________

____

____

2. ______________________________

____

____

Active Vocabulary

Review Vocabulary Exponential Function

Define *exponential function* in your own words. (Lesson 3-1)

exponential function ▶ ______________________________

____

Define *logarithmic function* in your own words. (Lesson 3-2)

logarithmic function ▶ ______________________________

____

New Vocabulary Fill in the blank with the correct term.

linearize A ______________ models growth that was initially exponential but that slows down and levels out, approaching a horizontal asymptote.

logistic growth function To __________ data would be to transform the data so that it appears to cluster about a line.

Main Idea	Details

Exponential, Logarithmic, and Logistic Modeling
pp. 200–204

Sketch a graph for each model.

Linear Regression	Quadratic Regression	Power Regression
Exponential Regression	Logarithmic Regression	Logistic Regression

Linearizing Data
pp. 204–206

Linearize the data according to the power model and write an equation of a line of best fit. Then use this linear model for the transformed data to find a model for the original data.

x	1	2	3	4	5	6
y	2	16	54	128	250	432

____

____

NAME ______________________ DATE ____________ PERIOD ____________

CHAPTER 3

Exponential and Logarithmic Functions

Tie It Together

Complete the first and third columns with one or more details. Complete the middle column with a description of the relationship between the rectangular boxes.

Exponential Function	← →	Logarithmic Function
Shape of Graph	← →	Shape of Graph
Solving Exponential Equations	← →	Solving Logarithmic Equations
$f(x) = e^x$	← →	$f(x) = \ln x$

NAME ______________________________ DATE ____________ PERIOD ____________

CHAPTER 3 Exponential and Logarithmic Functions

Before the Test

Now that you have read and worked through the chapter, think about what you have learned and complete the table below. Compare your previous answers with these.

1. Write an **A** if you agree with the statement.

2. Write a **D** if you disagree with the statement.

Exponential and Logarithmic Functions	After You Read
• The function $f(x) = x^3$ is an exponential function.	
• A natural logarithm is a logarithm with base e.	
• $\log_b x + \log_b y = \log_b (x + y)$	
• $36^{x+1} = 6^{(x+1)^2}$	
• Data exhibiting rapid growth or decay can be modeled using a logarithmic function.	

Math Online Visit *glencoe.com* to access your textbook, more examples, self-check quizzes, personal tutors, and practice tests to help you study for concepts in Chapter 3.

Are You Ready for the Chapter Test?

Use this checklist to help you study.

- ☐ I completed the Chapter 3 Study Guide and Review in the textbook.
- ☐ I took the Chapter 3 Practice Test in the textbook.
- ☐ I used the online resources for additional review options.
- ☐ I reviewed my homework assignments and made corrections to incorrect answers.
- ☐ I reviewed all vocabulary terms from the chapter.

Study Tips

- Review information every day to keep it fresh in your mind and to help reduce the amount of studying before test day.
- Look over your notes, and review your corrected homework. If you have any questions about any of the concepts, ask your teacher before the day of the test.

NAME ______________________ DATE __________ PERIOD __________

CHAPTER 4

Trigonometric Functions

Before You Read

Before you read the chapter, think about what you know about trigonometric functions. List three things that you already know about them in the first column. Then list three things that you would like to learn about them in the second column.

K What I know ...	W What I want to find out ...

Note-Taking Tips

- **Do not write every word. Concentrate on the main ideas and concepts.**
  Draw and label diagrams that correspond to the main ideas and concepts in your notes.
- **Always write clear and concise notes so that they can be easily read when studying for a quiz or exam.**
  Read your notes later in the day and rewrite any parts that are confusing.

NAME ______________________ DATE __________ PERIOD __________

CHAPTER 4

Trigonometric Functions

Key Points

Scan the pages in the chapter. Write at least one specific fact concerning each lesson. For example, in the lesson on right triangle trigonometry, one fact might be that the cosecant, secant, and cotangent functions are reciprocal functions of the sine, cosine, and tangent functions, respectively. After completing the chapter, you can use this table to review for your chapter test.

Lesson	Fact
4-1 Right Triangle Trigonometry	
4-2 Degrees and Radians	
4-3 Trigonometric Functions on the Unit Circle	
4-4 Graphing Sine and Cosine Functions	
4-5 Graphing Other Trigonometric Functions	
4-6 Inverse Trigonometric Functions	
4-7 The Law of Sines and the Law of Cosines	

NAME ______________________ DATE ____________ PERIOD ____________

4-1 Right Triangle Trigonometry

Lesson 4-1

What You'll Learn

Scan the examples for Lesson 4-1. Predict two things that you think you will learn about right triangle trigonometry.

1. ____

____

____

2. ____

____

____

Active Vocabulary

New Vocabulary Fill in each blank with the correct term.

cosine If θ is an acute angle and the sine of θ is x, then the ____________ of x is the measure of angle θ.

inverse cosine Let θ be an acute angle in a right triangle and the abbreviations opp, adj, and hyp refer to the lengths of the side opposite θ, the side adjacent to θ, and the hypotenuse, respectively. Then ____________ $(\theta) = \frac{\text{adj}}{\text{hyp}}$.

inverse sine If θ is an acute angle and the tangent of θ is x, then the ____________ of x is the measure of angle θ.

inverse tangent Let θ be an acute angle in a right triangle and the abbreviations opp, adj, and hyp refer to the lengths of the side opposite θ, the side adjacent to θ, and the hypotenuse, respectively. Then ____________ $(\theta) = \frac{\text{opp}}{\text{adj}}$.

sine If θ is an acute angle and the cosine of θ is x, then the ____________ of x is the measure of angle θ.

tangent Let θ be an acute angle in a right triangle and the abbreviations opp, adj, and hyp refer to the lengths of the side opposite θ, the side adjacent to θ, and the hypotenuse, respectively. Then ____________ $(\theta) = \frac{\text{opp}}{\text{hyp}}$.

Main Idea	**Details**

Values of Trigonometric Ratios
pp. 220–222

Find the exact values of the six trigonometric functions of θ.

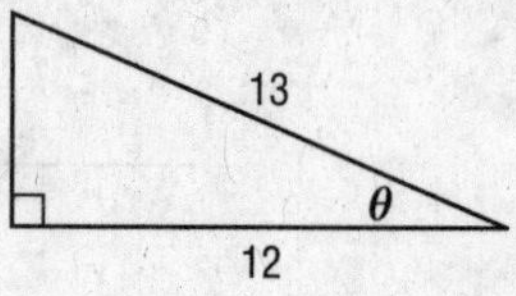

$\sin \theta =$ ______ $\cos \theta =$ ______ $\tan \theta =$ ______

$\csc \theta =$ ______ $\sec \theta =$ ______ $\cot \theta =$ ______

Solving Right Triangles
pp. 222–226

Solve $\triangle ABC$. Round side lengths to the nearest tenth and angle measures to the nearest degree.

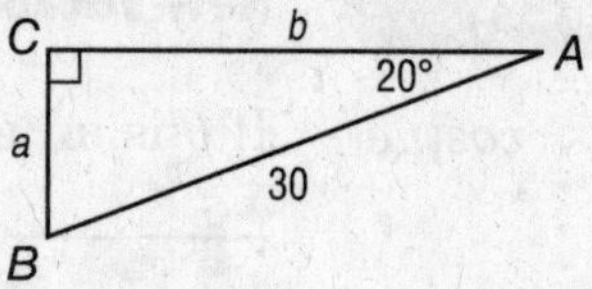

$B =$ __________ $a \approx$ __________ $b \approx$ __________

Helping You Remember

In your own words, describe the relationship that exists between the sine and inverse sine functions. Be specific.

____

____

____

____

____

____

NAME ______________________ DATE ____________ PERIOD ____________

4-2 Degrees and Radians

What You'll Learn

Scan the text under the *Now* heading. List two things that you will learn in the lesson.

1. ______________________________

____

2. ______________________________

____

Active Vocabulary

New Vocabulary Label the diagram with the terms listed at the left.

coterminal angles

initial side

standard position

terminal side

vertex

Positive ____________

in ____________

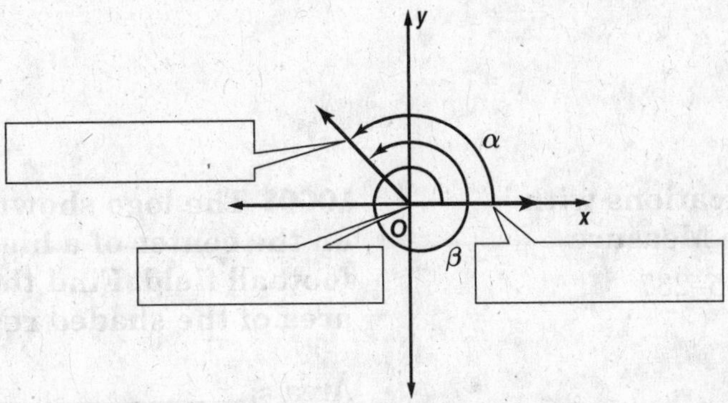

Vocabulary Link *Sector* is a word that is used in everyday English. Find the definition of *sector* using a dictionary. Explain how the English definition can help you remember how *sector* is used in mathematics.

____

____

____

____

____

Lesson 4-2

Main Idea	Details
Angles and Their Measure pp. 231–234	**Identify all angles that are coterminal with 30°. Then find and draw one positive and one negative angle coterminal with 30°.**
Applications with Angle Measure pp. 235–237	**LOGOS The logo shown is painted on the center of a high school football field. Find the approximate area of the shaded region.** Area ≈ ____________

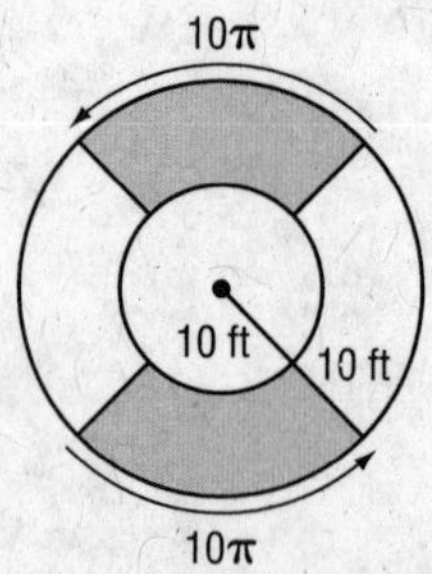

Helping You Remember

Suppose θ is a central angle in a circle of radius r. Explain how to find the length of the intercepted arc s. Include a drawing with your explanation.

NAME ______________________ DATE __________ PERIOD __________

4-3 Trigonometric Functions on the Unit Circle

What You'll Learn

Scan the text in Lesson 4-3. Write two facts that you learned about reference angles.

1. ______________________

____

____

2. ______________________

____

Active Vocabulary

New Vocabulary Write the definition next to each term.

circular function ▶ ______________________

____

period ▶ ______________________

____

periodic function ▶ ______________________

____

quadrantal angle ▶ ______________________

____

reference angle ▶ ______________________

____

unit circle ▶ ______________________

Lesson 4-3

Main Idea | **Details**

Trigonometric Functions of Any Angle
pp. 242–247

Complete the table.

θ	$30°$ or $\frac{\pi}{6}$	$45°$ or $\frac{\pi}{4}$	$60°$ or $\frac{\pi}{3}$
$\sin\theta$			
$\cos\theta$			
$\tan\theta$			

Trigonometric Functions on the Unit Circle
pp. 247–250

Fill in the blanks to complete the unit circle. Then, use the unit circle to find the values.

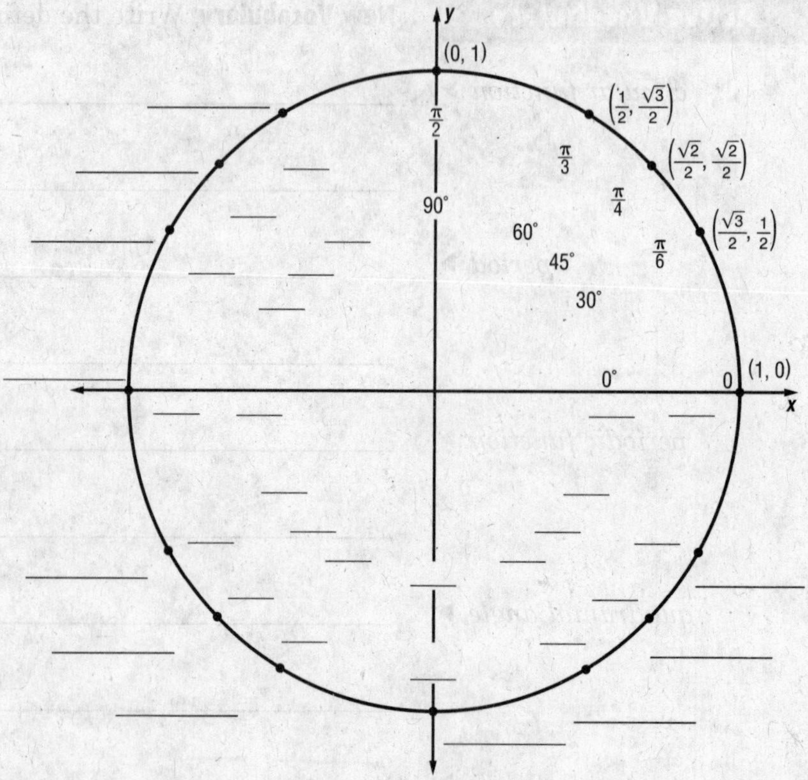

$\sin\frac{7\pi}{6} =$ ______ $\cos 300° =$ ______ $\tan\frac{2\pi}{3} =$ ______

$\cot 60° =$ ______ $\sec\frac{5\pi}{4} =$ ______ $\csc 150° =$ ______

NAME ______________________ DATE __________ PERIOD __________

4-4 Graphing Sine and Cosine Functions

What You'll Learn

Scan Lesson 4-4. Write two things that you already know about the sine and cosine functions.

1. ______________________________

____

2. ______________________________

____

Active Vocabulary

New Vocabulary Match the term with its definition by drawing a line to connect the two.

amplitude	any transformation of a sine function
frequency	half the distance between the maximum and minimum values of a sinusoidal function
midline	the distance between any two sets of repeating points on the graph of a sinusoidal function
period	the number of cycles a sinusoidal function completes in a one unit interval
phase shift	the difference between the horizontal position of a sinusoidal function and that of an otherwise similar sinusoidal function
sinusoid	a vertical translation of a sinusoidal function
vertical shift	the reference line about which a sinusoidal function oscillates

Main Idea	Details

Transformations of Sine and Cosine Functions

pp. 256–262

State the amplitude, period, frequency, phase shift, and vertical shift of $y = 2 \cos (x - 3\pi) + 4$.

Amplitude: ____________ Period: ____________

Frequency: ____________ Phase shift: ____________

Vertical shift: ________

Applications of Sinusoidal Functions

p. 263

METEOROLOGY The average monthly temperatures for Chicago, Illinois, are shown.

Month	Jan	Feb	Mar	Apr	May	June
Temp. (°F)	22	27	37.3	47.8	58.7	68.2
Month	July	Aug	Sep	Oct	Nov	Dec
Temp. (°F)	73.3	71.7	63.8	52.1	39.3	27.4

Source: U.S. National Oceanic and Atmospheric Administration

a. Write a function that models the monthly temperatures, using $x = 1$ to represent January.

Step 1 Make a scatter plot of the data and choose a model.

____

Step 2 Use the maximum and minimum values of the data to find a, b, c, and d.

a = ________ b = ________

c = ________ d = ________

Step 3 Write the function using the values for a, b, c, and d.

____

b. According to your model, what is Chicago's average monthly temperature in May?

____

NAME ______________________ DATE ____________ PERIOD ____________

4-5 Graphing Other Trigonometric Functions

What You'll Learn

Scan Lesson 4-5. Predict two things that you expect to learn based on the headings and Key Concept boxes.

1. ______________________________

____

____

2. ______________________________

____

____

Active Vocabulary

New Vocabulary Write the correct term next to each definition.

______________ an object is in this type of motion when the amplitude is determined by the function $y = ke^{-ct}$

______________ the reduction in the amplitude of a sinusoidal wave

______________ a function of the form $y = f(x) \sin bx$ or $y = f(x) \cos bx$, where $f(x)$ is the damping factor

______________ the resulting wave when the amplitude of a sinusoidal function is reduced

______________ the function $f(x)$ in a damped trigonometric function of the form $y = f(x) \sin bx$ or $y = f(x) \cos bx$

Review Vocabulary Define *sinusoid* in your own words. (Lesson 4-4)

sinusoid ______________________________

____

Main Idea	Details
Tangent and Reciprocal Functions pp. 269–274	**Locate the vertical asymptotes, and sketch the graph of $y = \tan 2x - 1$.**

Vertical asymptotes: ______________________

Graph:

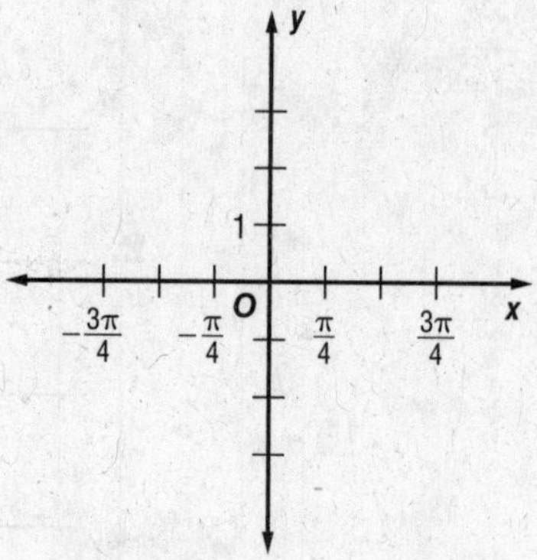

Damped Trigonometric Functions pp. 275–276	**Identify the damping factor $f(x)$ of each function. Then describe the behavior of the graph.**

1. $y = \frac{2}{3}x \sin x$

____

____

2. $y = (x + 3)^2 \cos x$

____

____

3. $y = 4^x \sin x$

____

____

Helping You Remember

Explain the relationship between the cosine and secant, the sine and cosecant, and the tangent and cotangent functions.

____

____

4-6 Inverse Trigonometric Functions

Lesson 4-6

What You'll Learn

Scan Lesson 4-6. List two headings that you would use to make an outline of this lesson.

1. ______________________

____

2. ______________________

____

Active Vocabulary

Review Vocabulary Define *function* in your own words. (Lesson 1-1)

function ▶ ______________________

____

Define *inverse function* in your own words. (Lesson 1-7)

inverse function ▶ ______________________

____

New Vocabulary Write the definition next to each term.

arcsine function ▶ ______________________

____

arccosine function ▶ ______________________

____

arctangent function ▶ ______________________

____

Main Idea	Details

Inverse Trigonometric Functions
pp. 280–285

Complete each table.

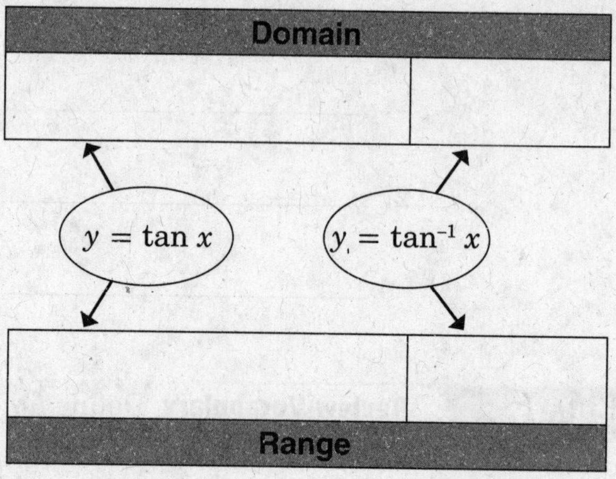

Compositions of Trigonometric Functions
pp. 286–287

Find the exact value of each expression, if it exists.

1. $\cos\left(\cos^{-1}\frac{1}{2}\right) =$ ____________

2. $\sin^{-1}\left(\cos\frac{\pi}{4}\right) =$ ____________

3. $\cos\left(\tan^{-1}\frac{-\sqrt{3}}{3}\right) =$ ____________

4. $\tan\left(\sin^{-1} -\frac{\sqrt{3}}{2}\right) =$ ____________

Helping You Remember

Explain the relationship that exists between the sine function and the inverse sine function. Provide two examples.

____

____

____

NAME ______________________ DATE __________ PERIOD __________

4-7 The Law of Sines and the Law of Cosines

What You'll Learn

Scan Lesson 4-7. Predict two things that you expect to learn based on the headings and Key Concept boxes.

1. ______________________________

____

____

2. ______________________________

____

____

Active Vocabulary

New Vocabulary Match the term with its definition by drawing a line to connect the two.

ambiguous case	a formula used to solve an oblique triangle when given the measures of three sides or the measures of two sides and the included angle
Heron's Formula	triangles that are not right triangles
Law of Cosines	a formula used to solve an oblique triangle when given the measures of two angles and a nonincluded side, two angles and the included side, or two sides and a nonincluded angle
Law of Sines	a formula of finding the area of a triangle when the measures of all three sides are known
oblique triangles	while solving an oblique triangle when given the measures of two sides and a nonincluded angle there may be no solution, one solution, or two solutions

Lesson 4-7

Main Idea	Details

Solve Oblique Triangles
pp. 291–296

Identify the law that you would use to solve an oblique triangle with the given measures.

Oblique Triangles	
Known Measures	**Law of Sines or Law of Cosines**
two sides and a nonincluded angle	
two angles and the included side	
two sides and the included angle	
two angles and a nonincluded side	
three sides	

Find Areas of Oblique Triangles
pp. 296–297

Use Heron's Formula to find the area of each triangle. Round to the nearest tenth, if necessary.

1. $a = 5$ in., $b = 12$ in., $c = 13$ in. ____________

2. $a = 17$ cm, $b = 21$ cm, $c = 30$ cm ____________

3. $a = 25$ yd, $b = 32$ yd, $c = 43$ yd ____________

Helping You Remember

In your own words, explain how to find the area of any triangle when you know the measure of two sides and the included angle. Use a diagram to illustrate.

____

____

____

NAME ______________________ DATE ____________ PERIOD ____________

CHAPTER 4 Trigonometric Functions

Tie It Together

Complete the table.

Function	Domain	Range	Graph	Period
$y = \sin x$			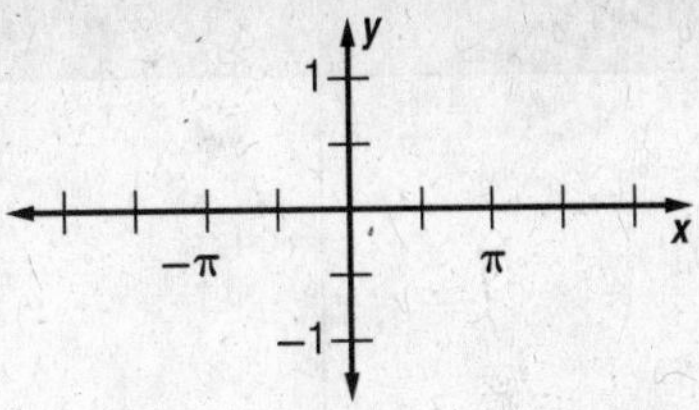	
$y = \sin^{-1} x$			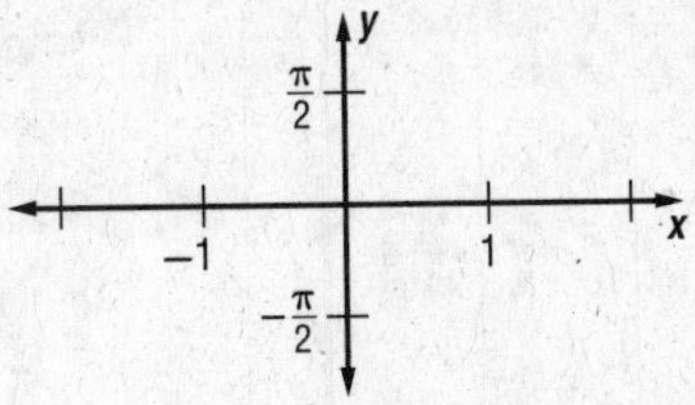	
$y = \cos x$			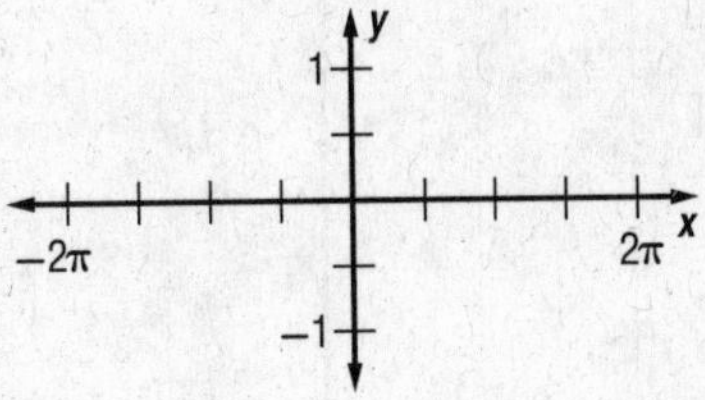	
$y = \cos^{-1} x$			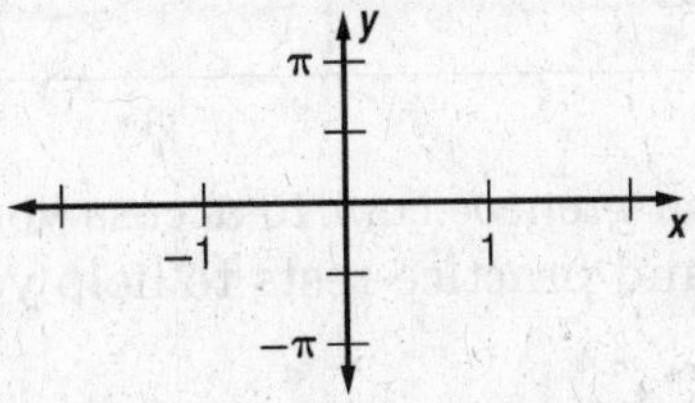	
$y = \tan x$			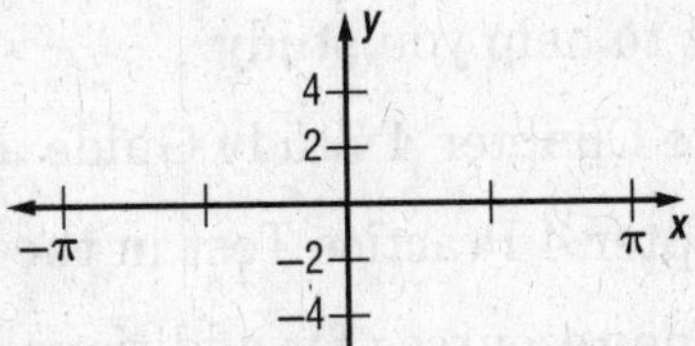	
$y = \tan^{-1} x$			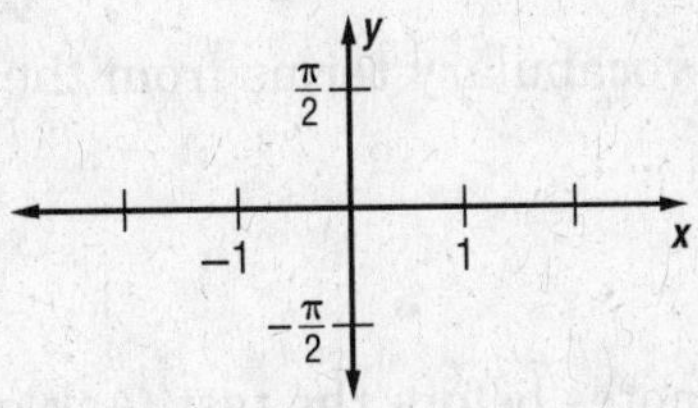	

NAME ______________________ DATE __________ PERIOD __________

CHAPTER 4

Before the Test

Review the ideas that you listed in the table at the beginning of the chapter. Cross out any incorrect information in the first column. Then complete the table by filling in the third column.

K What I know...	W What I want to find out...	L What I Learned...

Math Online Visit *glencoe.com* to access your textbook, more examples, self-check quizzes, personal tutors, and practice tests to help you study for concepts in Chapter 4.

Are You Ready for the Chapter Test?

Use this checklist to help you study.

- ☐ I completed the Chapter 4 Study Guide and Review in the textbook.
- ☐ I took the Chapter 4 Practice Test in the textbook.
- ☐ I used the online resources for additional review options.
- ☐ I reviewed my homework assignments and made corrections to incorrect answers.
- ☐ I reviewed all vocabulary terms from the chapter.

Study Tip

- Review your notes before the test. Ask questions if you need any topics clarified.

NAME ______________________ DATE ____________ PERIOD ____________

CHAPTER 5

Trigonometric Identities and Equations

Before You Read

Before you read the chapter, think about what you know about trigonometric identities and equations. List three things that you already know about them in the first column. Then list three things that you would like to learn about them in the second column.

K What I know ...	W What I want to find out ...

Note-Taking Tips

- **Write questions that you have about the lesson in the margin of your notes.** Record the answers to your questions as you work through the lesson.
- **Include graphs and charts that present the information introduced in a format that is easy to read and study.**

NAME ______________________ DATE ____________ PERIOD ____________

CHAPTER 5 Trigonometric Identities and Equations

Key Points

Scan the pages in the chapter. Write at least one specific fact concerning each lesson. For example, in the lesson on trigonometric identities, one fact might be that a function f is a cofunction of a function g if $f(\alpha) = g(\beta)$ when α and β are complementary angles. After completing the chapter, you can use this table to review for your chapter test.

Lesson	Fact
5-1 Trigonometric Identities	
5-2 Verifying Trigonometric Identities	
5-3 Solving Trigonometric Equations	
5-4 Sum and Difference Identities	
5-5 Multiple-Angle and Product-to-Sum Identities	

NAME ______________________ DATE ____________ PERIOD ____________

5-1 Trigonometric Identities

What You'll Learn

Scan the text under the *Now* heading. List two things that you will learn in the lesson.

1. ____

____

____

2. ____

____

____

Active Vocabulary

Review Vocabulary Complete each identity. (Lesson 4-1)

$\frac{1}{\csc \theta} =$ ______________ $\frac{1}{\sec \theta} =$ ______________

$\frac{1}{\tan \theta} =$ ______________ $\frac{\sin \theta}{\cos \theta} =$ ______________

$\frac{\cos \theta}{\sin \theta} =$ ______________ $\frac{1}{\cos \theta} =$ ______________

New Vocabulary Fill in each blank with the correct term.

identity A function f is a(n) ______________ of a function g if $f(\alpha) = g(\beta)$ when α and β are complementary angles.

trigonometric identity A(n) ______________ is an equation in which the left side is equal to the right side for all values of the variable for which both sides are defined.

cofunction A(n) ______________ is an identity involving trigonometric functions.

Lesson 5-1

Main Idea	Details
Basic Trigonometric Identities pp. 312–314	**Complete each of the following charts.**

List the Pythagorean identities.

__________ __________ __________

List the cofunction identities.

__________ __________ __________

__________ __________ __________

List the odd-even identities.

__________ __________ __________

__________ __________ __________

Simplify and Rewrite Trigonometric Expressions
pp. 315–316

Simplify $\sin x - \cos\left(\frac{\pi}{2} - x\right)\cos^2 x$.

__________	Original equation
= __________	Cofunction Identity
= __________	Factor $\sin x$ from each term.
= __________	Pythagorean Identity
= __________	Simplify.

Helping You Remember

Explain how the unit circle could be used to remember the odd-even identities. Include a diagram illustrating your explanation.

____

____

____

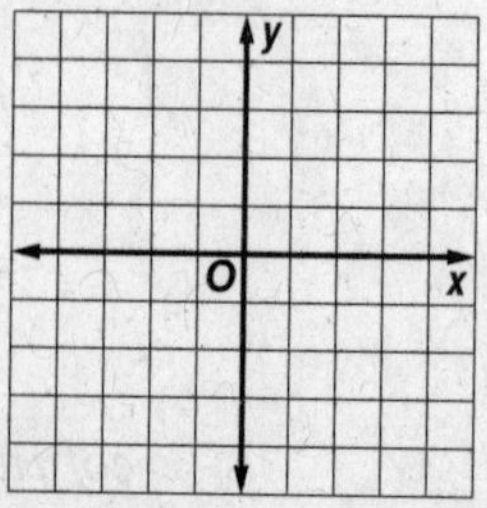

NAME ______________________ DATE ____________ PERIOD ____________

5-2 Verifying Trigonometric Identities

What You'll Learn

Scan the Examples for Lesson 5-2. Predict two things that you think you will learn about verifying trigonometric identities.

1. ______________________

____

____

2. ______________________

____

____

Active Vocabulary

New Vocabulary Write the definition next to the term.

verify an identity ______________________

____

____

____

Vocabulary Link *Verify* is a word that is used in everyday English. Use a dictionary to find a definition of the word *verify*. Explain how the English definition can help you remember what it means to *verify an identity* in mathematics.

____

____

____

____

Lesson 5-2

Main Idea	Details
Verify Trigonometric Identities pp. 320–323	**Complete each of the following steps to verify** $\frac{\sin x}{1 + \cos x} + \frac{\sin x}{1 - \cos x} = 2 \csc x$.

______________________ Original identity

______________________ Start with the left-hand side.

______________________ Common denominator

______________________ Combine like terms.

______________________ Pythagorean Identity

______________________ Reciprocal Identity

Identify Identities and Nonidentities p. 323	**Use your graphing calculator to test whether** $\sec^2 x \tan^2 x - 1 = \tan^2 x$ **is an identity. If it appears to be an identity, verify it. If it is not, find an** x**-value for which both sides are defined but not equal.**

Graph left side: Graph right side:

[–3.14, 3.14] scl: 1.57 by [–3, 3] scl: 1 [–3.14, 3.14] scl: 1.57 by [–3, 3] scl: 1

Is this an identity? ____________

x-value: ________________

NAME ______________________ DATE ____________ PERIOD ____________

5-3 Solving Trigonometric Equations

What You'll Learn

Scan Lesson 5-3. Predict two things that you expect to learn based on the headings and examples.

1. ______________________

2. ______________________

Active Vocabulary

Review Vocabulary Define each term in your own words.

identity ▶ ______________________

trigonometric identity ▶ ______________________

verify an identity ▶ ______________________

cofunction ▶ ______________________

Lesson 5-3

Main Idea	Details

Use Algebraic Techniques to Solve
pp. 327–329

Complete each step to solve $6 \tan x + \sqrt{3} = 3 \tan x$.

______________________ Original equation

______________________ Subtract $3 \tan x$ from each side.

______________________ Subtract $\sqrt{3}$ from each side.

______________________ Divide each side by 3.

______________________ Find solution(s).

Use Trigonometric Identities to Solve
p. 330

Complete each of the following steps to solve $2 \sin^2 x + \cos x = 1$ in the interval $[0, 2\pi]$.

______________________ Original equation

______________________ Pythagorean Identity

______________________ Distributive Property

______________________ Subtract 1 from each side.

______________________ Multiply by –1.

______________________ Factor.

______________________ Solve.

______________________ Find solution(s).

Helping You Remember

To help you remember the algebraic techniques that are used in solving trigonometric equations, write examples that involve factoring a quadratic trinomial and factoring the difference of two squares.

____

NAME ______________________ DATE ____________ PERIOD ____________

5-4 Sum and Difference Identities

What You'll Learn

Scan the lesson. Write two things that you already know about trigonometric functions.

1. ____

____

____

2. ____

____

____

Active Vocabulary

Review Vocabulary Find each value. (Lesson 4-3)

$\sin \frac{\pi}{3} =$ ________ $\tan \frac{5\pi}{6} =$ ________

$\cos \pi =$ ________ $\cot \frac{\pi}{2} =$ ________

$\csc \frac{7\pi}{4} =$ ________ $\cos \frac{4\pi}{3} =$ ________

$\sin \frac{3\pi}{4} =$ ________ $\tan \frac{\pi}{6} =$ ________

$\sec \frac{\pi}{6} =$ ________ $\csc \frac{2\pi}{3} =$ ________

New Vocabulary Write the definition of *reduction identity.*

reduction identity ____

____

____

Lesson 5-4

Main Idea	Details
Evaluate Trigonometric Functions pp. 336–340	**Find the exact value of $\cos \frac{5\pi}{12}$.** $\cos \frac{5\pi}{12} = \cos$ ______________ $\frac{\pi}{6} + \frac{\pi}{4} = \frac{5\pi}{12}$ $=$ ______________ Cosine Sum Identity $=$ ______________ Substitute. $=$ ______________ Multiply and combine like terms.
Solve Trigonometric Equations p. 340	**Find the solution to each expression in the interval $[0, 2\pi)$. Show your work.** **1.** $\sin\left(\frac{\pi}{6} + x\right) + \sin\left(\frac{\pi}{6} - x\right) = \frac{\sqrt{2}}{2}$ ______________ ______________ ______________ ______________ **2.** $\tan(\pi + x) + \tan(\pi + x) = 2\sqrt{3}$ ______________ ______________ ______________ ______________

NAME ______________________ DATE __________ PERIOD __________

5-5 Multiple-Angle and Product-to-Sum Identities

What You'll Learn

Scan Lesson 5-5. List two headings that you would use to make an outline of this lesson.

1. ______________________________

____

2. ______________________________

____

Active Vocabulary

Review Vocabulary Complete each identity. (Lesson 5-1)

$\sin(-\theta) =$ __________ $\cos(-\theta) =$ __________

$\tan(-\theta) =$ __________ $\csc(-\theta) =$ __________

$\sec(-\theta) =$ __________ $\cot(-\theta) =$ __________

New Vocabulary Complete each identity.

sine double angle identity $\sin(2\theta) =$ __________

cosine double angle identity $\cos(2\theta) =$ __________

tangent double angle identity $\tan(2\theta) =$ __________

sine half-angle identity $\sin\left(\frac{\theta}{2}\right) =$ __________

cosine half-angle identity $\cos\left(\frac{\theta}{2}\right) =$ __________

tangent half-angle identity $\tan\left(\frac{\theta}{2}\right) =$ __________

Lesson 5-5

Main Idea	Details

Use Multiple-Angle Identities
pp. 346–349

Complete each of the following steps. If $\sin \theta = \frac{5}{13}$ in the interval $\left(\frac{\pi}{2}, \pi\right)$, find $\sin 2\theta$, $\cos 2\theta$, and $\tan 2\theta$.

$\sin 2\theta =$ ____________ Sine Double-Angle Identity

$=$ ____________ $\sin \theta = \frac{5}{13}$, $\cos \theta = \frac{12}{13}$

$=$ ______ Simplify.

$\cos 2\theta =$ ____________ Cosine Double-Angle Identity

$=$ ____________ $\cos \theta = \frac{12}{13}$, $\sin \theta = \frac{5}{13}$

$=$ ______ Simplify.

$\tan 2\theta =$ ____________ Tangent Double-Angle Identity

$=$ ____________ $\tan \theta = \frac{5}{12}$

$=$ ____________ Simplify.

Use Product-to-Sum Identities
pp. 350–351

Rewrite each product as a sum or difference.

1. $\cos 2x \sin 4x =$ ____________________

2. $\sin 5\theta \sin 3\theta =$ ____________________

Find each exact value.

3. $\sin 50° + \sin 40° =$ ____________

4. $\cos \frac{\pi}{8} - \cos \frac{3\pi}{8} =$ ____________

NAME ______________________ DATE __________ PERIOD __________

CHAPTER 5

Trigonometric Identities and Equations

Tie It Together

Complete the graphic organizer.

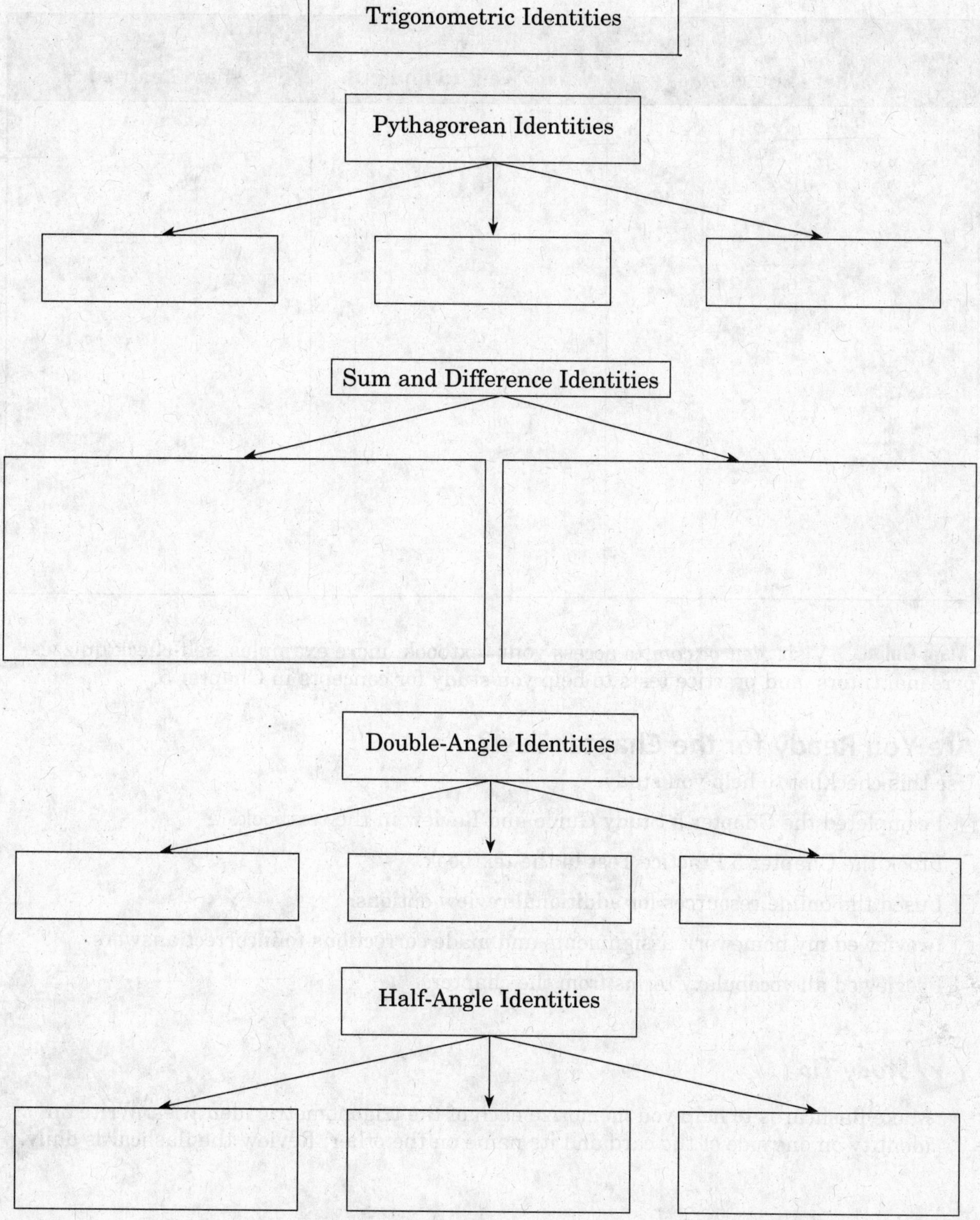

NAME ______________________ DATE ____________ PERIOD ____________

CHAPTER 5

Trigonometric Identities and Equations

Before the Test

Review the ideas that you listed in the table at the beginning of the chapter. Cross out any incorrect information in the first column. Then complete the third column of the table.

K What I know...	W What I want to find out...	L What I Learned...

Math Online Visit *glencoe.com* to access your textbook, more examples, self-check quizzes, personal tutors, and practice tests to help you study for concepts in Chapter 5.

Are You Ready for the Chapter Test?

Use this checklist to help you study.

- ☐ I completed the Chapter 5 Study Guide and Review in the textbook.
- ☐ I took the Chapter 5 Practice Test in the textbook.
- ☐ I used the online resources for additional review options.
- ☐ I reviewed my homework assignments and made corrections to incorrect answers.
- ☐ I reviewed all vocabulary terms from the chapter.

Study Tip

- Make flashcards to help you memorize each of the trigonometric identities. Write an identity on one side of the card and its name on the other. Review the flashcards daily.

NAME ______________________ DATE ____________ PERIOD ____________

CHAPTER 6 Systems of Equations and Matrices

Before You Read

Before you read the chapter, respond to these statements.

1. Write an **A** if you agree with the statement.

2. Write a **D** if you disagree with the statement.

Before You Read	Systems of Equations and Matrices
	• A multivariable linear system is a system of linear equations in more than two variables.
	• Matrix multiplication is a commutative operation.
	• Cramer's Rule uses inverse matrices to solve square systems of linear equations.
	• When finding the partial fraction decomposition of an improper rational expression, you should first divide the numerator by the denominator using polynomial division.
	• If a linear programming problem can be optimized, it occurs at an interior point of the region representing the set of feasible solutions.

Note-Taking Tips

- **Before each lesson, scan through the lesson. Write any questions that come to mind in your notes.**

  As you work through the lesson, record the answers to your questions. If you are unable to answer all of your questions, ask your teacher for help.

- **You may wish to use a highlighting marker to emphasize important concepts in your notes.**

NAME ______________________ DATE ____________ PERIOD ________

CHAPTER 6

Systems of Equations and Matrices

Key Points

Scan the pages in the chapter. Write at least one specific fact concerning each lesson. For example, in the lesson on linear optimization, one fact might be that there may be multiple solutions to a linear programming problem. After completing the chapter, you can use this table to review for your chapter test.

Lesson	Fact
6-1 Multivariable Linear Systems and Row Operations	
6-2 Matrix Multiplication, Inverses, and Determinants	
6-3 Solving Linear Systems Using Inverses and Cramer's Rule	
6-4 Partial Fractions	
6-5 Linear Optimization	

NAME ____________________ DATE ____________ PERIOD ________

6-1 Multivariable Linear Systems and Row Operations

Lesson 6-1

What You'll Learn

Scan the examples for Lesson 6-1. Predict two things that you think you will learn about solving systems of linear equations.

1. ______________________________

____

____

2. ______________________________

____

____

Active Vocabulary

New Vocabulary Label each diagram with a term listed at the left.

augmented matrix

coefficient matrix

elementary row operations

multivariable linear system

row-echelon form

$$\left.\begin{aligned} x - 2y + 3z &= 7 \\ 2x + y + z &= 4 \\ -3x + 2y - 2z &= -10 \end{aligned}\right\} \leftarrow \boxed{}$$

$\boxed{}$ ↓ $\boxed{}$ ↓

$$\left[\begin{array}{ccc|c} 1 & -2 & 3 & 7 \\ 2 & 1 & 1 & 4 \\ -3 & 2 & -2 & -10 \end{array}\right] \qquad \left[\begin{array}{ccc} 1 & -2 & 3 \\ 2 & 1 & 1 \\ -3 & 2 & -2 \end{array}\right]$$

$$\begin{array}{r} \frac{1}{2}R_1 \longrightarrow \\ -2R_1 + R_2 \longrightarrow \\ \end{array} \left[\begin{array}{ccc|c} \frac{1}{2} & -1 & \frac{3}{2} & \frac{7}{2} \\ 0 & 5 & -5 & -10 \\ -3 & 2 & -2 & -10 \end{array}\right]$$

↑ $\boxed{}$

Lesson 6-1 *(continued)*

Main Idea	Details

Gaussian Elimination
pp. 364–368

Perform the indicated elementary row operations to produce the row-echelon form of the given system of linear equations.

$$x - y + 2z = 6$$
$$2x + y + 5z = -9$$
$$y - 2z = -5$$

Augmented matrix $\left[\begin{array}{ccc|c} 1 & -1 & 2 & 6 \\ 2 & 1 & -5 & -9 \\ 0 & 1 & -2 & -5 \end{array}\right]$ $\frac{1}{3}R_2 \rightarrow \left[\begin{array}{ccc|c} 1 & -1 & 2 & 6 \\ \_\_ & \_\_ & \_\_ & \_\_ \\ 0 & 1 & -2 & -5 \end{array}\right]$

$R_2 - 2R_1 \rightarrow \left[\begin{array}{ccc|c} 1 & -1 & 2 & 6 \\ \_\_ & \_\_ & \_\_ & \_\_ \\ 0 & 1 & -2 & -5 \end{array}\right]$ $R_3 - R_2 \rightarrow \left[\begin{array}{ccc|c} 1 & -1 & 2 & 6 \\ 0 & 1 & -3 & -7 \\ \_\_ & \_\_ & \_\_ & \_\_ \end{array}\right]$

Gauss-Jordan Elimination
pp. 369–371

Solve the system of equations using Gauss-Jordan elimination.

$$x - y + z = 1$$
$$2x + y - z = -4$$
$$-x + 2y + 3z = 7$$

Write the augmented matrix.

____

Apply elementary row operations to obtain a row-echelon form.

____

Apply elementary row operations to obtain zeros above the leading 1s in each row.

____

____________________ Write the solution to the system.

6-2 Matrix Multiplication, Inverses, and Determinants

What You'll Learn

Scan the text under the *Now* heading. List two things that you will learn about in the lesson.

1. ______________________________

____

____

2. ______________________________

____

____

Active Vocabulary

New Vocabulary Fill in each blank with the correct term.

determinant The multiplicative identity for a set of square matrices is called the ____________.

identity matrix The multiplicative inverse of a square matrix is called its ____________ matrix.

inverse A matrix that has an inverse is said to be ____________.

invertible A(n) ____________ is a matrix that does not have an inverse.

singular matrix Let A be the matrix $\begin{bmatrix} a & b \\ c & d \end{bmatrix}$. The number $ad - cb$ is called the ____________ of the matrix A.

Main Idea	Details

Multiply Matrices
pp. 375–378

Use the following matrices to illustrate the Associative Property of Matrix Multiplication.

$$A = \begin{bmatrix} -1 & 3 \\ 0 & -4 \end{bmatrix}, B = \begin{bmatrix} 2 & -1 \\ 1 & -3 \end{bmatrix}, C = \begin{bmatrix} -5 & 1 \\ -3 & 0 \end{bmatrix}$$

$$(AB)C = A(BC)$$

$$\left(\begin{bmatrix} \_\_ & \_\_ \\ \_\_ & \_\_ \end{bmatrix}\begin{bmatrix} \_\_ & \_\_ \\ \_\_ & \_\_ \end{bmatrix}\right)\begin{bmatrix} \_\_ & \_\_ \\ \_\_ & \_\_ \end{bmatrix} = \begin{bmatrix} \_\_ & \_\_ \\ \_\_ & \_\_ \end{bmatrix}\left(\begin{bmatrix} \_\_ & \_\_ \\ \_\_ & \_\_ \end{bmatrix}\begin{bmatrix} \_\_ & \_\_ \\ \_\_ & \_\_ \end{bmatrix}\right)$$

$$\begin{bmatrix} \_\_ & \_\_ \\ \_\_ & \_\_ \end{bmatrix}\begin{bmatrix} -5 & 1 \\ -3 & 0 \end{bmatrix} = \begin{bmatrix} -1 & 3 \\ 0 & -4 \end{bmatrix}\begin{bmatrix} \_\_ & \_\_ \\ \_\_ & \_\_ \end{bmatrix}$$

$$\begin{bmatrix} \_\_ & \_\_ \\ \_\_ & \_\_ \end{bmatrix} = \begin{bmatrix} \_\_ & \_\_ \\ \_\_ & \_\_ \end{bmatrix}$$

Inverses and Determinants
pp. 379–382

Find the determinant and inverse of $\begin{bmatrix} -7 & 4 \\ 5 & -3 \end{bmatrix}$.

Determinant: ________

Inverse: ________

Helping You Remember

Identify which of the properties listed below apply to matrices. If a property does not apply, provide a counterexample.

Associative Property of Multiplication
Commutative Property of Multiplication
Distributive Property

____

____

____

____

6-3 Solving Linear Systems Using Inverses and Cramer's Rule

What You'll Learn

Scan the lesson. Write two things that you already know about solving systems of linear equations.

1. ______________________

2. ______________________

Active Vocabulary

Review Vocabulary Define *augmented matrix* in your own words. (Lesson 6-1)

____

Define *inverse matrix* in your own words. (Lesson 6-2)

____

New Vocabulary Write the definition next to each term.

Cramer's Rule ▶ ______________________

square system ▶ ______________________

Main Idea	Details
Use Inverse Matrices pp. 388–389	**Use an inverse matrix to solve the system of equations.**

$$-2x + 5y = 17$$
$$3x - 7y = -24$$

Write the system in matrix form $AX = B$.

____

Find A^{-1}.

$A^{-1} =$ ______________

Multiply A^{-1} by B.

$X =$ ______________

Use Cramer's Rule pp. 390–391	**Use Cramer's Rule to find the solution of the system of linear equations, if a unique solution exists.**

$$2x - 3y = -7$$
$$x + 4y = 2$$

Calculate the determinant of the coefficient matrix.

____

Solve for x.

____

Solve for y.

____

Write the solution to the system.

____

NAME _______________ DATE _______ PERIOD _______

6-4 Partial Fractions

What You'll Learn

Scan Lesson 6-4. List two headings that you would use to make an outline of this lesson.

1. ______________________________

____

____

2. ______________________________

____

____

Active Vocabulary

Review Vocabulary Define *polynomial function* in your own words. (Lesson 2-2)

____

____

____

Define *rational function* in your own words. (Lesson 2-5)

____

____

____

New Vocabulary Write the correct term next to each definition.

______________ ► one of the rational expressions in the sum of two or more rational expressions that represents a rational function

______________ ► the sum of the rational expressions that represent a rational function

Main Idea	Details

Linear Factors
pp. 398–400

Find the partial fraction decomposition of $\frac{8x - 7}{x^2 - x - 2}$.

Form of partial fraction decomposition ____________________

Multiply each side by the LCD. ____________________

Distributive Property ____________________

Group like terms. ____________________

Equate the coefficients to obtain a system of two equations. ____________________

Solve the system. ____________________

Write the partial decomposition. ____________________

Irreducible Quadratic Factors
pp. 401

Write the appropriate form of the partial fraction decomposition for each rational expression. Do not solve for the variables.

1. $\frac{4x^2 - x + 8}{x^3 + 4x}$ ____________________

2. $\frac{2x^3 + 5x^2 + 2x + 9}{(x + 2)^2(x^2 - 7)}$ ____________________

Helping You Remember

A classmate told you that the appropriate form of the partial fraction decomposition of $\frac{2x^2 + 7x + 3}{(x + 3)^2}$ is $\frac{A}{(x + 3)^2}$. Explain your classmate's error. Then give the appropriate form of the decomposition.

____

____

____

NAME ______________________ DATE __________ PERIOD __________

6-5 Linear Optimization

What You'll Learn

Scan Lesson 6-5. Predict two things that you expect to learn based on the headings and Key Concept boxes.

1. ______________________________

____

____

2. ______________________________

____

____

Active Vocabulary

New Vocabulary Match the term with its definition by drawing a line to connect the two.

constraints	the process of finding a minimum or maximum value for a specific quantity
feasible solutions	the process of finding a minimum or maximum value for a function for a region defined by linear inequalities
linear programming	a function to be optimized in a two-dimensional linear programming problem
objective function	a system of linear inequalities in a two-dimensional linear programming problem
optimization	the solution set of a system of linear inequalities in a two-dimensional linear programming problem

Main Idea	Details
Linear Programming pp. 405–407	**Complete each statement to show the steps to follow when solving a linear programming problem.**

Step 1: Graph ______________________________

____

____

Step 2: Find ______________________________

____

Step 3: Evaluate ______________________________

____

____

____

No or Multiple Optimal Solutions
pp. 408–409

Fill in the blank with the correct word or phrase.

1. If the graph of the objective function f to be optimized is coincident with one side of the region of feasible solutions, f has ____________________.

2. If the region does not form a polygon, but is instead ____________________, the objective function f may have no minimum or maximum value.

Helping You Remember

Provide an example of a linear system that has multiple optimal solutions and an example of a linear system that is unbounded.

Multiple Optimal Solutions	Unbounded
______________________________	______________________________
______________________________	______________________________

NAME ______________________ DATE ____________ PERIOD ________

CHAPTER 6

Systems of Equations and Matrices

Tie It Together

Use the following system of equations to complete the organizers.

$2x + 4y + 5z = 5$
$x + 3y + 3z = 2$
$x + 2y + 2z = 1$

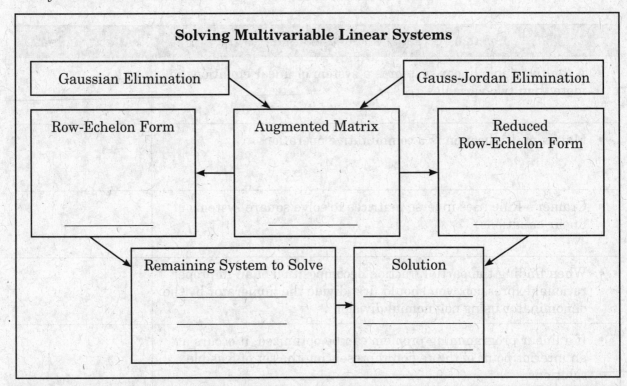

Solving Multivariable Linear Systems

Gaussian Elimination → Augmented Matrix ← Gauss-Jordan Elimination

Row-Echelon Form ← Augmented Matrix → Reduced Row-Echelon Form

Row-Echelon Form ____________

Augmented Matrix ____________

Reduced Row-Echelon Form ____________

Row-Echelon Form → Remaining System to Solve → Solution ← Reduced Row-Echelon Form

Remaining System to Solve
____
____
____

Solution
____
____

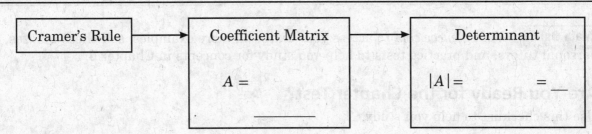

Cramer's Rule → Coefficient Matrix → Determinant

Coefficient Matrix: $A =$ ____________

Determinant: $|A| =$ ____________ $=$ __

Apply Cramer's Rule. $|A_x| =$ ____________ $=$ __, $|A_y| =$ ____________ $=$ __, $|A_z| =$ ____________ $=$ __

Solve. $x = \frac{|A_x|}{|A|} =$ ____________, $y = \frac{|A_y|}{|A|} =$ ____________, $z = \frac{|A_z|}{|A|} =$ ____________

NAME ______________________ DATE ____________ PERIOD ____________

CHAPTER 6

Systems of Equations and Matrices

Before the Test

Now that you have read and worked through the chapter, think about what you have learned. Complete the table below. Compare your previous answers with these.

1. Write an **A** if you agree with the statement.
2. Write a **D** if you disagree with the statement.

Systems of Equations and Matrices	After You Read
• A multivariable linear system is a system of linear equations in more than two variables.	
• Matrix multiplication is a commutative operation.	
• Cramer's Rule uses inverse matrices to solve square systems of linear equations.	
• When finding the partial fraction decomposition of an improper rational expression, you should first divide the numerator by the denominator using polynomial division.	
• If a linear programming problem can be optimized, it occurs at an interior point of the region representing the set of feasible solutions.	

Math Online Visit *glencoe.com* to access your textbook, more examples, self-check quizzes, personal tutors, and practice tests to help you study for concepts in Chapter 6.

Are You Ready for the Chapter Test?

Use this checklist to help you study.

☐ I completed the Chapter 6 Study Guide and Review in the textbook.

☐ I took the Chapter 6 Practice Test in the textbook.

☐ I used the online resources for additional review options.

☐ I reviewed my homework assignments and made corrections to incorrect answers.

☐ I reviewed all vocabulary terms from the chapter.

Study Tips

- Designate a distraction-free place to study at home. Establish a routine by studying around the same time each day.

NAME ______________________ DATE ____________ PERIOD ____________

CHAPTER 7

Conic Sections and Parametric Equations

Before You Read

Before you read the chapter, respond to these statements.

1. Write an **A** if you agree with the statement.
2. Write a **D** if you disagree with the statement.

Before You Read	Conic Sections and Parametric Equations
	• The equation of a parabola that opens upward or downward is $(y - k)^2 = 4p(x - h)$.
	• The foci of an ellipse lie on the minor axis.
	• A hyperbola has two asymptotes.
	• The equation $6x^2 - 4xy - 2y^2 + 7x + 2y - 5 = 0$ represents a rotated hyperbola.
	• $2x + 6y = 5$ and $x - 4y = 2$ are parametric equations.

Note-Taking Tips

- **Be sure to include the standard forms of the equations of all of the conic sections in your notes. Indicate what each variable in the equations represents.**

  Draw and label a diagram of each figure.

- **Write notes to help you remember important concepts.**

  For example, make notes about a^2 being the greater number in the denominator of the equation for an ellipse.

NAME ______________________ DATE ____________ PERIOD ____________

CHAPTER 7 Conic Sections and Parametric Equations

Key Points

Scan the pages in the chapter. Write at least one specific fact concerning each lesson. For example, in the lesson on parabolas, one fact might be that the four common conic sections are the parabola, the ellipse, the circle, and the hyperbola. After completing the chapter, you can use this table to review for your chapter test.

Lesson	Fact
7-1 Parabolas	
7-2 Ellipses and Circles	
7-3 Hyperbolas	
7-4 Rotations of Conic Sections	
7-5 Parametric Equations	

NAME ______________________ DATE __________ PERIOD __________

7-1 Parabolas

What You'll Learn

Scan the examples for Lesson 7-1. Predict two things that you think you will learn about parabolas.

1. ______________________________

____

____

2. ______________________________

____

____

Active Vocabulary

New Vocabulary Match the term with its definition by drawing a line to connect the two.

axis of symmetry	the intersection of a parabola and its axis of symmetry
conic section	the fixed point from which the locus of points on a parabola are equidistant
directrix	the locus of points in a plane that are equidistant from a fixed point and a specific line
focus	a figure formed when a plane intersects a double-napped right cone
parabola	the line from which points on a parabola are equidistant
vertex	a line perpendicular to the directrix through the focus of a parabola

Lesson 7-1

Main Idea	Details

Analyze and Graph Parabolas
pp. 422–425

For $(y - 6)^2 = 20(x - 8)$, identify the vertex, focus, axis of symmetry, and directrix. Then graph the parabola.

vertex: ______________

focus: ______________

axis of symmetry: ______________

directrix: ______________

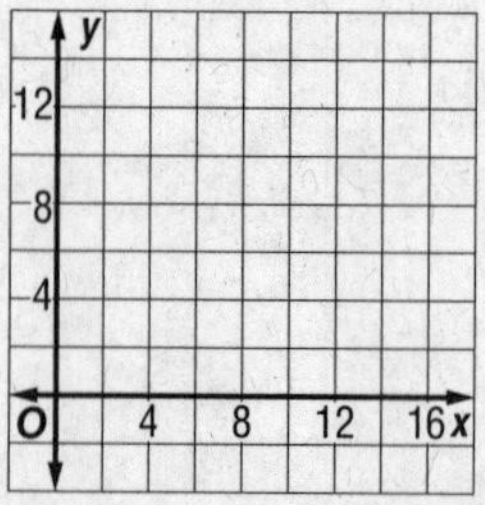

Equations of Parabolas
pp. 425–427

Write an equation for and graph the parabola with focus (3, 1) that opens upward and contains the point (5, 1). Complete the steps below.

______________________________ Standard form

______________________________ Substitute.

______________________________ Simplify.

______________________________ Solve for p.

____

______________________________ Write the equation.

Graph:

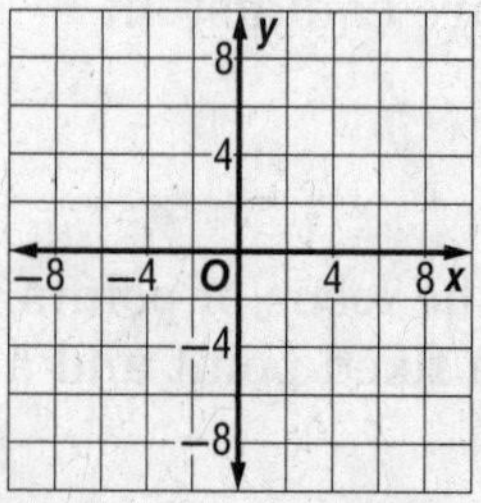

Helping You Remember

Explain how to remember that $(x - h)^2 = 4p(y - k)$ is a vertical parabola while $(y - k)^2 = 4p(x - h)$ is a horizontal parabola.

____

____

____

NAME ______________________ DATE ____________ PERIOD ____________

7-2 Ellipses and Circles

What You'll Learn

Scan the text in Lesson 7-2. Write two facts that you learned about ellipses and circles.

1. ____

____

____

2. ____

____

____

Active Vocabulary

New Vocabulary Label the diagram with the correct terms.

center ______________________

co-vertex

ellipse

focus

major axis

minor axis

vertex

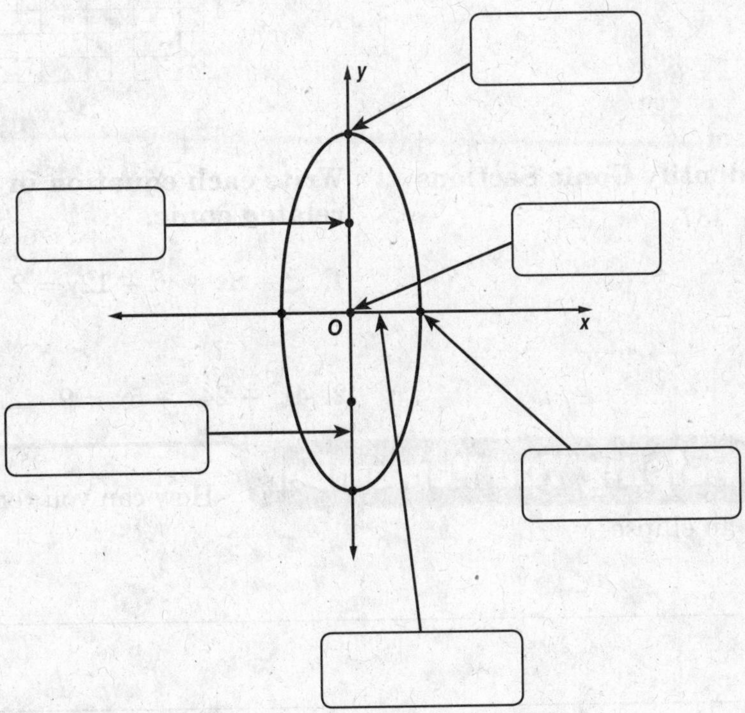

Lesson 7-2

Main Idea | **Details**

Analyze and Graph Ellipses and Circles
pp. 432–436

Graph the ellipse given by $\frac{(x-4)^2}{16} + \frac{(y+2)^2}{9} = 1$.

orientation: _______________

center: _______________

foci: _______________

vertices: _______________

co-vertices: _______________

major axis: _______________

minor axis: _______________

graph:

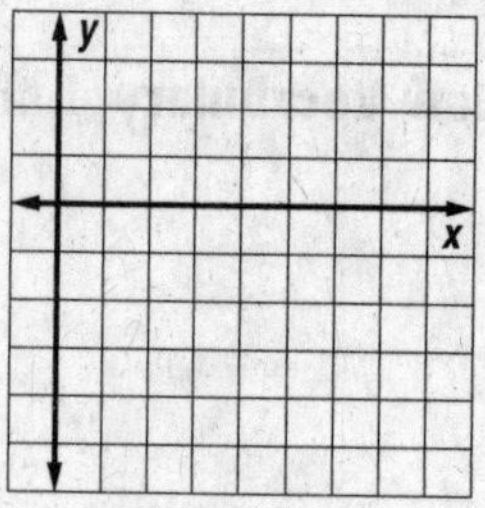

Identify Conic Sections
p. 437

Write each equation in standard form. Identify the related conic.

1. $x^2 + 8x + y^2 + 12y = 2$ _______________

2. $4x^2 + 24x + 5y = 9$ _______________

Helping You Remember

How can you remember that $a^2 = b^2 + c^2$ for an ellipse?

____

____

____

7-3 Hyperbolas

What You'll Learn

Scan Lesson 7-3. Predict two things that you expect to learn based on the headings and Key Concept box.

1. ______________________

2. ______________________

Active Vocabulary

Review Vocabulary Define each term in your own words.

parabola ▶ ______________________
(Lesson 7-1)

ellipse ▶ ______________________
(Lesson 7-2)

New Vocabulary Write the definition next to each term.

conjugate axis ▶ ______________________

transverse axis ▶ ______________________

hyperbola ▶ ______________________

NAME ______________________________ DATE ____________ PERIOD ____________

Lesson 7-3 *(continued)*

Main Idea	Details

Analyze and Graph Hyperbolas
pp. 442–447

Graph the hyperbola given by $\frac{(x-1)^2}{4} - \frac{(y-5)^2}{9} = 1$.

$a =$ ________ $b =$ ________ $c =$ ________

orientation: ______________________

center: ______________ vertices: ______________

foci: ______________________

asymptotes: ______________________

graph:

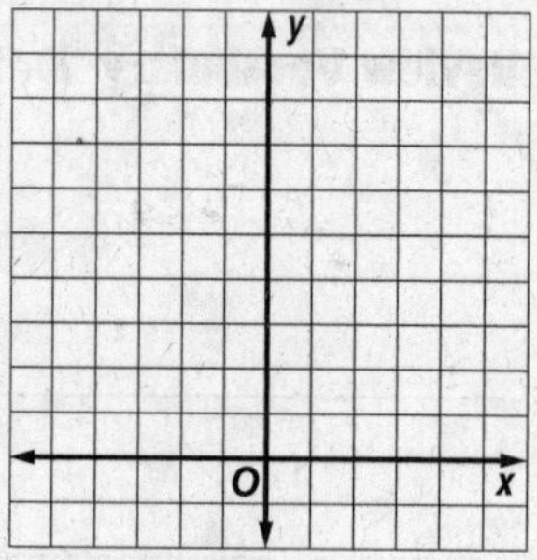

Identify Conic Sections
pp. 447–448

Use the discriminant to identify each conic section.

1. $2x^2 + 3y^2 - 5x + 4y = 7$ ____________
2. $4x^2 - 10x + 4y^2 + 2y = 9$ ____________
3. $9x^2 + 6xy + y^2 - 3x + 8y = 11$ ____________
4. $7x^2 - 3xy - 4y^2 + 18x + y = 2$ ____________

Helping You Remember

How can you distinguish between a and b in the equation of an ellipse or a hyperbola?

____

____

____

NAME ______________________ DATE ____________ PERIOD ____________

7-4 Rotations of Conic Sections

What You'll Learn

Scan the lesson. Write two things that you already know about conic sections.

1. ______________________

2. ______________________

Active Vocabulary

Review Vocabulary Use each term in a sentence that shows its mathematical meaning.

conic section ▶ (Lesson 7-1) ______________________

conjugate axis ▶ (Lesson 7-3) ______________________

degenerate conic ▶ (Lesson 7-1) ______________________

eccentricity ▶ (Lesson 7-2) ______________________

transverse axis ▶ (Lesson 7-3) ______________________

NAME ______________________ DATE ____________ PERIOD ____________

Main Idea	**Details**

Rotations of Conic Sections
pp. 454–457

Write the standard form equation of $x^2 - 4xy + y^2 = 6$ in the $x'y'$-plane for $\theta = 45°$. Then identify the conic.

Rotation equations for x and y: ______________________

____

Substitute.

$x =$ ______________ $y =$ ______________

Substitute into the original equation.

____

Simplify.

____

____

____

Identify the conic. ______________

Graph Rotated Conics
pp. 457–458

Graph $(x' - 3)\ 2 = 6(y' + 1)$ if it has been rotated 30° from its position in the xy-plane.

$x =$ ______________ $y =$ ______________

vertex:

____

axis of symmetry:

____

graph:

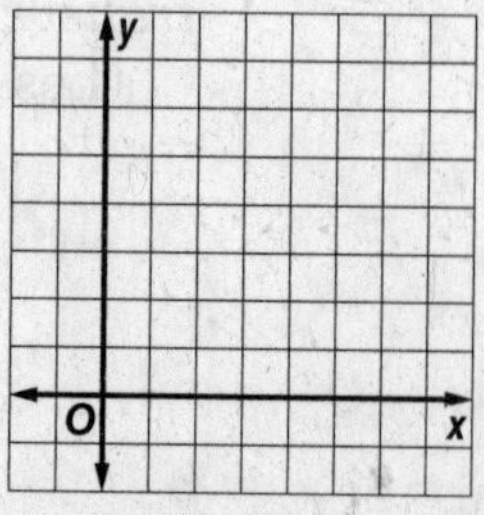

NAME ______________________ DATE __________ PERIOD __________

7-5 Parametric Equations

What You'll Learn

Scan Lesson 7-5. List two headings that you would use to make an outline of this lesson.

1. ______________________________

____

2. ______________________________

____

Active Vocabulary

New Vocabulary Write the correct term next to each definition.

______________ ▶ the direction of a curve when points are plotted in the order of increasing values of t

______________ ▶ an arbitrary value, such as a time or an angle measurement

______________ ▶ equations that can express the position of an object as a function of time

______________ ▶ the form of the equation of a graph using a pair of parametric equations

______________ ▶ the set of ordered pairs $(f(t), g(t))$, where f and g are continuous functions

______________ ▶ the form of an equation in terms of (x, y) coordinates

Main Idea	Details

Graph Parametric Equations
pp. 464–467

Write each pair of parametric equations in rectangular form.

1. $y = t^2 + 3$
$x = t - 1$ ______________________

2. $y = 2t$
$x = t^2 + 4$ ______________________

3. $x = 2 \cos t$
$y = 2 \sin t$ ______________________

4. $y = 5 \cos t$
$x = 3 \sin t$ ______________________

Projectile Motion
pp. 467–468

BASEBALL Davis hit a baseball from a height of 3 feet above the ground at an angle of 24° with the horizontal at an initial velocity of 110 feet per second. How far will the ball travel before it hits the ground?

Write parametric equations for the flight of the ball.

____

____

Find the maximum height of the ball.

____

The ball will travel ____________ horizontally before it hits the ground.

Helping You Remember

In the second problem above, explain why the 3 and $-16t^2$ terms are both in the y-equation and not in the x-equation.

____

____

____

NAME ______________________ DATE __________ PERIOD __________

CHAPTER 7

Conic Sections and Parametric Equations

Tie It Together

Complete the graphic organizer.

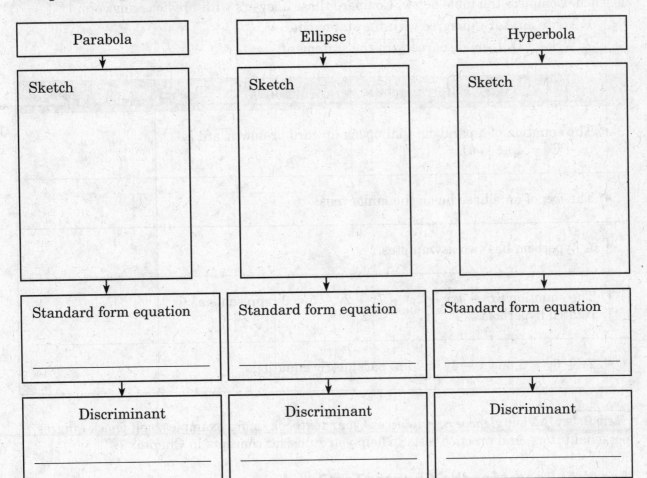

Parabola	Ellipse	Hyperbola
Sketch	Sketch	Sketch
Standard form equation ______	Standard form equation ______	Standard form equation ______
Discriminant ______	Discriminant ______	Discriminant ______

Rotation Equations

$x =$ ______ $x' =$ ______

$y =$ ______ $y' =$ ______

Projectile Motion

Horizontal Distance: ______

Vertical Position: ______

NAME ______________________ DATE ____________ PERIOD ____________

CHAPTER 7 Conic Sections and Parametric Equations

Before the Test

Now that you have read and worked through the chapter, think about what you have learned. Complete the table below. Compare these answers with previous answers.

1. Write an **A** if you agree with the statement.
2. Write a **D** if you disagree with the statement.

Conic Sections and Parametric Equations	After You Read
• The equation of a parabola that opens upward or downward is $(y - k)^2 = 4p(x - h)$.	
• The foci of an ellipse lie on the minor axis.	
• A hyperbola has two asymptotes.	
• The equation $6x^2 - 4xy - 2y^2 + 7x + 2y - 5 = 0$ represents a rotated hyperbola.	
• $2x + 6y = 5$ and $x - 4y = 2$ are parametric equations.	

Math Online Visit *glencoe.com* to access your textbook, more examples, self-check quizzes, personal tutors, and practice tests to help you study for concepts in Chapter 7.

Are You Ready for the Chapter Test?

Use this checklist to help you study.

- ☐ I completed the Chapter 7 Study Guide and Review in the textbook.
- ☐ I took the Chapter 7 Practice Test in the textbook.
- ☐ I used the online resources for additional review options.
- ☐ I reviewed my homework assignments and made corrections to incorrect answers.
- ☐ I reviewed all vocabulary terms from the chapter.

Study Tips

- Review the standard form of the equations for each conic every day.
- Practice drawing a graph of each conic and labeling all of the important points and lines on the figure.

NAME _______________ DATE _______ PERIOD _______

CHAPTER 8 Vectors

Before You Read

Before you read the chapter, think about what you know about vectors. List three things that you already know about them in the first column. Then list three things that you would like to learn about them in the second column.

K What I know...	W What I want to find out...

Note-Taking Tips

- **Make annotations.**
  Annotations are usually notes taken in the margins of books that you own to organize the text for review or study.
- **Be an active listener in class.**
  Take notes, circle or highlight information that your teacher stresses, and ask questions when ideas are unclear to you.

NAME ______________________ DATE ____________ PERIOD ____________

CHAPTER 8

Vectors

Key Points

Scan the pages in the chapter. Write at least one specific fact concerning each lesson. For example, in the lesson on geometric vectors, one fact might be that a vector with its initial point at the origin is said to be in standard position. After completing the chapter, you can use this table to review for your chapter test.

Lesson	Fact
8-1 Introduction to Vectors	
8-2 Vectors in the Coordinate Plane	
8-3 Dot Products and Vector Projections	
8-4 Vectors in Three-Dimensional Space	
8-5 Dot and Cross Products of Vectors in Space	

NAME ______________________ DATE __________ PERIOD __________

8-1 Introduction to Vectors

What You'll Learn

Scan Lesson 8-1. Predict two things that you expect to learn based on the headings and Key Concept boxes.

1. ______________________________

____

____

2. ______________________________

____

____

Active Vocabulary

New Vocabulary Label the diagram with the terms listed at the left.

equivalent vectors

initial point

opposite vectors

parallel vectors

terminal point

standard position

resultant

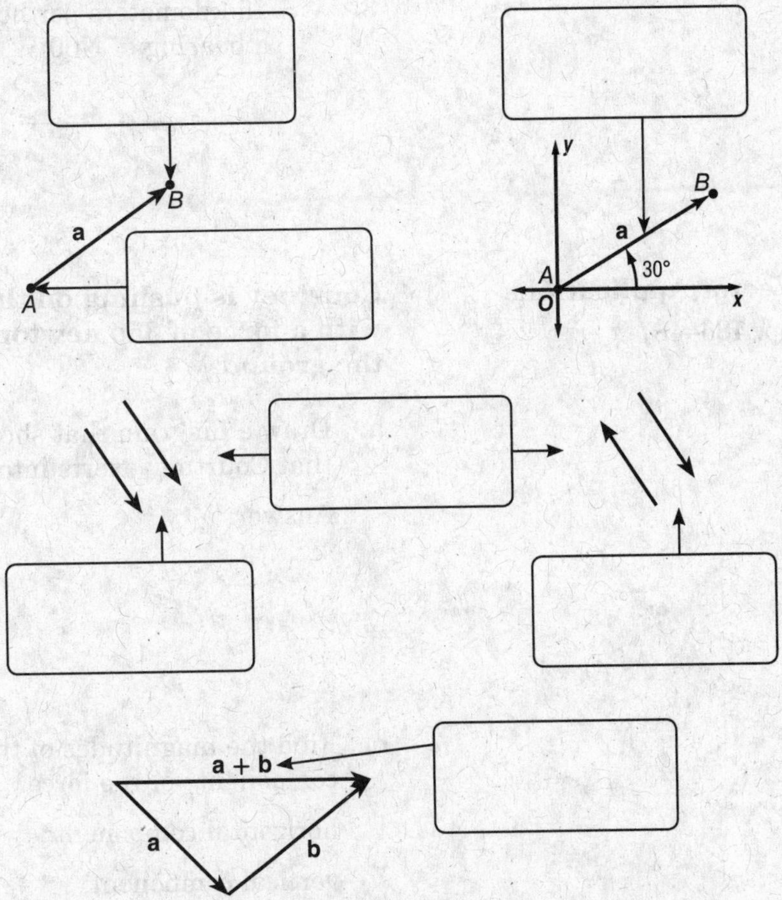

Lesson 8-1

Main Idea	**Details**

Vectors
pp. 482–485

Match each quantity described the corresponding arrow diagram that best represents the situation.

1. **u** = 15 kilometers per hour at a bearing of S40°E

   ____

A.

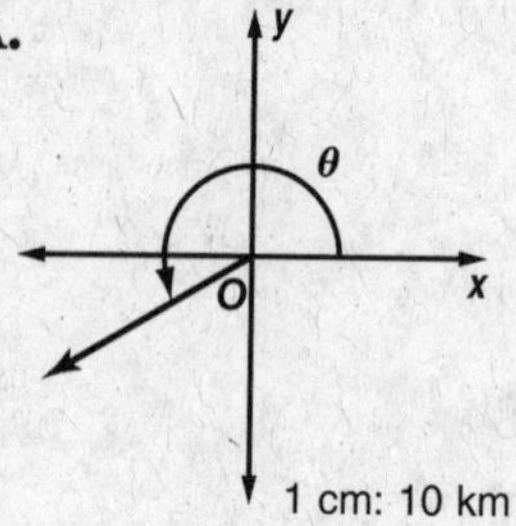

2. **w** = 15 kilometers per hour at 210° to the horizontal

   ____

B.

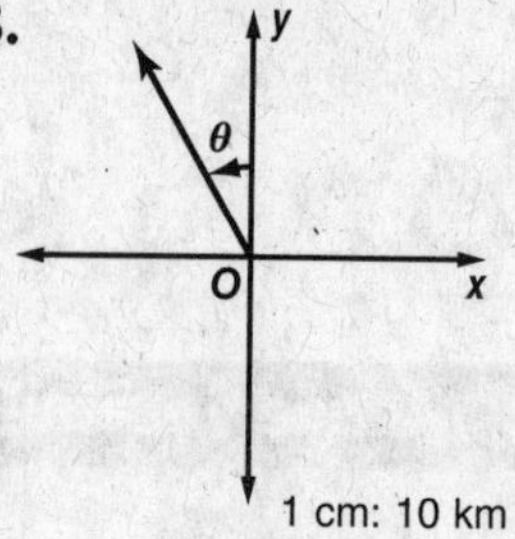

3. **v** = 15 kilometers per hour at a bearing of N60°W

   ____

C.

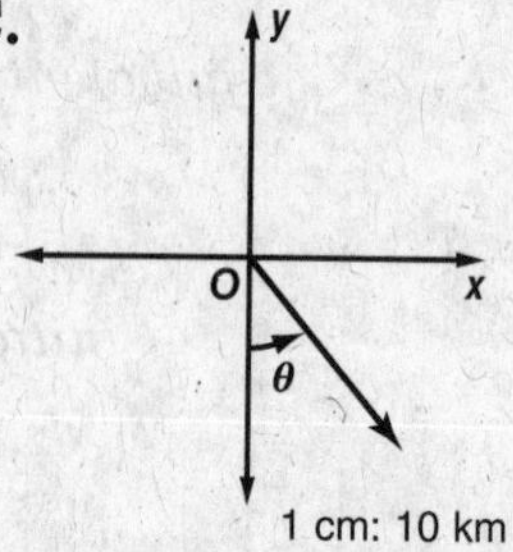

Vector Applications
pp. 486–487

Courtney is pushing on the handle of a snow shovel with a force of 350 newtons at an angle of 58° with the ground.

a. Draw a diagram that shows the resolution of the force that Courtney exerts into its rectangular components.

Answer:

b. Find the magnitudes of the horizontal and vertical components of the force.

horizontal component: ____________

vertical component: ____________

NAME ______________________ DATE ____________ PERIOD ____________

8-2 Vectors in the Coordinate Plane

What You'll Learn

Scan the examples for Lesson 8-2. Predict two things that you think you will learn about vectors in the coordinate plane.

1. ______________________________

____

2. ______________________________

____

Active Vocabulary

Review Vocabulary Define *vector* in your own words. (Lesson 8-1)

____

____

Define *standard position* in your own words. (Lesson 8-1)

____

____

Define *true bearing* in your own words. (Lesson 8-1)

____

____

New Vocabulary Fill in each blank with the correct term.

component form	A vector that has a magnitude of 1 unit is called a(n) ____________.
unit vector	The vector sum $a\mathbf{i} + b\mathbf{j}$ is called a(n) ____________ of the vectors $\mathbf{i}$ and $\mathbf{j}$.
linear combination	A way of describing a vector in standard position in the coordinate plane using the coordinates of the vector's terminal point is called the ____________ of the vector.

Lesson 8-2

Main Idea	Details
Vectors in the Coordinate Plane pp. 492–493	**Find each of the following for $\mathbf{u} = \langle 4, -1 \rangle$, $\mathbf{v} = \langle -2, 3 \rangle$, $\mathbf{w} = \langle -1, 0 \rangle$, and $k = -2$.**

1. $|\mathbf{u}| =$ ______________________
2. $|\mathbf{v}| =$ ______________________
3. $|\mathbf{w}| =$ ______________________
4. $\mathbf{u} + \mathbf{w} =$ ______________________
5. $k\mathbf{w} - \mathbf{v} =$ ______________________
6. $\mathbf{u} + k\mathbf{w} + \mathbf{v} =$ ______________________
7. $\frac{1}{|\mathbf{w}|}\mathbf{u} - \mathbf{v} =$ ______________________

Unit Vectors
pp. 494–496

Find the component form of the vector w with magnitude 6 and direction angle 60°. Then write w as a linear combination of i and j. Draw and label a diagram illustrating these relationships.

$\mathbf{w} =$ ______________________

$=$ ______________________

$=$ ______________________

Helping You Remember

What role does the inverse tangent play in finding the direction angle of a given vector? Illustrate with an example.

____

____

____

____

8-3 Dot Products and Vector Projections

What You'll Learn

Scan the text in Lesson 8-3. Write two facts that you learned about the dot product of vectors in a plane.

1. ______________________

____

____

2. ______________________

____

____

Active Vocabulary

Review Vocabulary Define *unit vector* in your own words. (Lesson 8-2)

____

____

Define *vector* in your own words. (Lesson 8-1)

____

____

New Vocabulary Match each term with its definition by drawing a line to connect the two.

dot product	the magnitude of a force applied to an object multiplied by the distance through which the object moves parallel to this applied force
orthogonal	perpendicular vectors with a dot product of zero
work	the expression $a_1b_1 + a_2b_2$ given $\mathbf{a} = \langle a_1, a_2\rangle$ and $\mathbf{b} = \langle b_1, b_2\rangle$

Main Idea	Details

Dot Product
pp. 500–502

Find the dot product of u and v. Then write *yes* or *no* to indicate whether u and v are orthogonal.

Vectors **u** and **v**	Dot Product	Orthogonal
1. $\mathbf{u} = \langle -2, 5\rangle, \mathbf{v} = \langle 4, -2\rangle$	______	______
2. $\mathbf{u} = \langle 3, 2\rangle, \mathbf{v} = \langle -2, 3\rangle$	______	______
3. $\mathbf{u} = \langle -6, -9\rangle, \mathbf{v} = \langle 3, -2\rangle$	______	______
4. $\mathbf{u} = \langle 0, 0\rangle, \mathbf{v} = \langle -5, 3\rangle$	______	______

Vector Projection
pp. 503–505

A 90-pound box is sitting at the top of a ramp that has an incline of 45°. Ignoring the force of friction, what force is required to keep the box from sliding down the ramp?

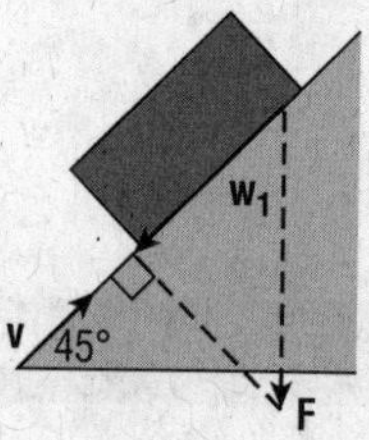

The force exerted due to gravity is

F = ____________________.

Step 1 Find a unit vector **v** in the direction of the ramp.

$$\mathbf{v} = \langle |\mathbf{v}| (\cos\theta), |\mathbf{v}| (\sin\theta)\rangle$$

= ____________________

= ____________________

Step 2 Find $\mathbf{w}_1$, the projection of **F** onto unit vector **v**.

$\text{proj}_{\mathbf{v}}\mathbf{F}$ = ____________________

= ____________________

= ____________________

= ____________________

The force required to keep the box from sliding down the ramp is ____________________.

NAME ______________________ DATE ____________ PERIOD ____________

8-4 Vectors in Three-Dimensional Space

What You'll Learn

Scan the text under the *Now* heading. List two things that you will learn about in the lesson.

1. ______________________________

____

2. ______________________________

____

Active Vocabulary

Review Vocabulary Define *rectangular components* in your own words. (Lesson 8-1)

____

____

____

New Vocabulary Write the definition next to each term.

ordered triple ▶ ______________________________

____

z-axis ▶ ______________________________

____

octants ▶ ______________________________

____

three-dimensional coordinate system ▶ ______________________________

____

____

Lesson 8-4

Main Idea	Details

Coordinates in Three Dimensions
pp. 510–511

Find the length and midpoint of the segment with the given endpoints.

1. (−1, 4, 7), (5, 6, −3)

   length: ____________ midpoint: ____________

2. (−2, 6, 1), (−6, −4, 7)

   length: ____________ midpoint: ____________

3. (3, −4, 8), (9, 2, −1)

   length: ____________ midpoint: ____________

4. (−5, −6, 1), (4, −7, 8)

   length: ____________ midpoint: ____________

Vectors in Space
pp. 512–513

Find each of the following for $\mathbf{x} = \langle -2, 5, -1 \rangle$, $\mathbf{y} = \langle 4, -3, 0 \rangle$, and $\mathbf{z} = \langle -2, -1, -3 \rangle$.

1. $2\mathbf{x} - \mathbf{y} =$ ____________

2. $\mathbf{z} - \mathbf{x} - \mathbf{y} =$ ____________

3. $-3\mathbf{y} + \mathbf{x} + 2\mathbf{z} =$ ____________

4. $4\mathbf{x} - 2\mathbf{y} + 3\mathbf{z} =$ ____________

Helping You Remember

Compare and contrast adding two vectors in a two-dimensional space to adding two vectors in a three-dimensional space, and provide an example for each.

____

____

____

____

____

8-5 Dot and Cross Products of Vectors in Space

What You'll Learn

Scan Lesson 8-5. List two headings that you would use to make an outline of this lesson.

1. ______________________

____

2. ______________________

____

Active Vocabulary

Review Vocabulary Define *three-dimensional coordinate system* in your own words. (Lesson 8-4)

____

____

____

New Vocabulary Write the correct term next to each definition.

______________ ▶ $\mathbf{t} \cdot (\mathbf{u} \times \mathbf{v}) = \begin{vmatrix} t_1 & t_2 & t_3 \\ u_1 & u_2 & u_3 \\ v_1 & v_2 & v_3 \end{vmatrix}$, where $\mathbf{t} = t_1\mathbf{i} + t_2\mathbf{j} + t_3\mathbf{k}$, $\mathbf{u} = u_1\mathbf{i} + u_2\mathbf{j} + u_3\mathbf{k}$, and $\mathbf{v} = v_1\mathbf{i} + v_2\mathbf{j} + v_3\mathbf{k}$

______________ ▶ the vector $\mathbf{a} \times \mathbf{b} = (a_2b_3 - a_3b_2)\mathbf{i} - (a_1b_3 - a_3b_1)\mathbf{j} + (a_1b_2 - a_2b_1)\mathbf{k}$, where $\mathbf{a} = a_1\mathbf{i} + a_2\mathbf{j} + a_3\mathbf{k}$ and $\mathbf{b} = b_1\mathbf{i} + b_2\mathbf{j} + b_3\mathbf{k}$

______________ ▶ a polyhedron with faces that are all parallelograms

______________ ▶ a vector quantity that measures how effectively a force applied to a lever causes rotation along the axis of rotation

Main Idea | **Details**

Dot Products in Space
p. 518

Find the angle θ between u and v if u = ⟨−2, 1, 3⟩ and v = ⟨−1, 4, 2⟩.

Dot Product

u **v** = ______________

= ______________

Magnitude of **u**

|**u**| = ______________

= ______________

Magnitude of **v**

|**v**| = ______________

= ______________

Find θ.

cos θ = ______________

θ = ______________

$\theta \approx$ ______________

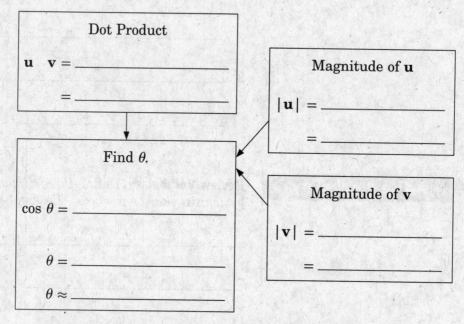

Cross Products
pp. 519–521

Find the cross product of a = ⟨−2, 3, 1⟩ and b = ⟨3, −1, 4⟩. Then show that a × b is orthogonal to both a and b.

a × b = ______ = ______ **i** − ______ **j** + ______ **k**

= ______**i** − ______**j** + ______**k** = ______________

(a × b) • a | **(a × b) • b**

______________ | ______________

______________ | ______________

______________ | ______________

____

____

NAME ______________________ DATE __________ PERIOD __________

CHAPTER 8

Vectors

Tie It Together

Complete the diagram that compares vectors in a two-dimensional space and vectors in a three-dimensional space.

Two-Dimensional Space $\mathbf{u} = \langle -1, -3\rangle$ and $\mathbf{v} = \langle 4, 2\rangle$	↔	**Three-Dimensional Space** $\mathbf{u} = \langle -1, -3, 4\rangle$ and $\mathbf{v} = \langle 4, 2, -1\rangle$
Graph of Vectors **u** and **v** 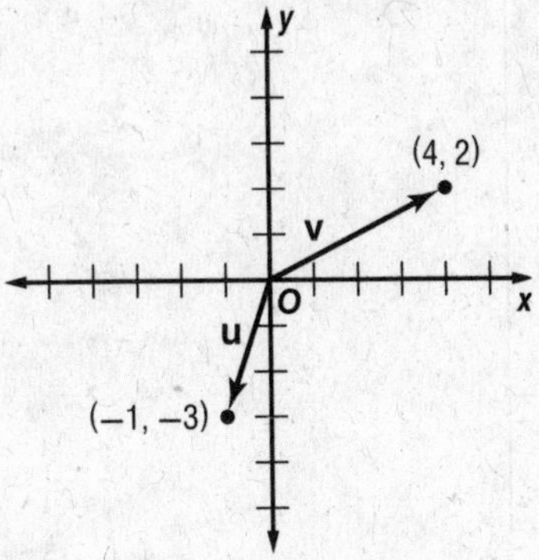	↔	Locate and Graph Vectors **u** and **v** 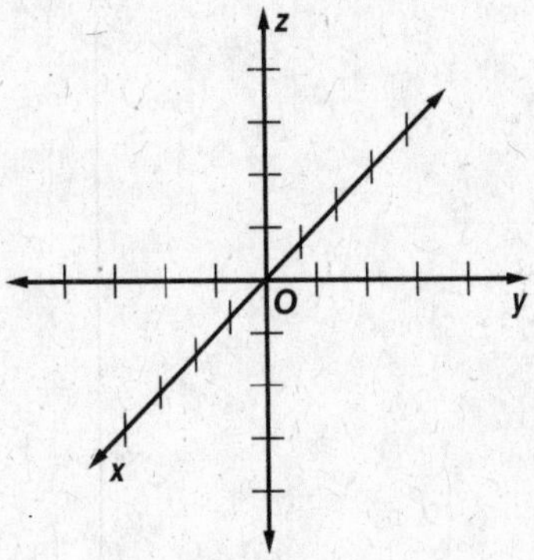
Magnitude $\lvert\mathbf{u}\rvert$ = ____________ $\lvert\mathbf{v}\rvert$ = ____________	↔	Magnitude $\lvert\mathbf{u}\rvert$ = ____________ $\lvert\mathbf{v}\rvert$ = ____________
Dot Product $\mathbf{u} \cdot \mathbf{v}$ = ____________ ____________	↔	Dot Product $\mathbf{u} \cdot \mathbf{v}$ = ____________ ____________
Angle Between **u** and **v** $\cos\theta = \dfrac{\mathbf{u} \cdot \mathbf{v}}{\lvert\mathbf{u}\rvert\ \lvert\mathbf{v}\rvert} =$ ____________ $\theta =$ ____________ $=$ ____________	↔	Angle Between **u** and **v** $\cos\theta = \dfrac{\mathbf{u} \cdot \mathbf{v}}{\lvert\mathbf{u}\rvert\ \lvert\mathbf{v}\rvert} =$ ____________ $\theta =$ ____________ $\approx$ ____________

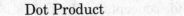

NAME ______________________ DATE ____________ PERIOD ____________

CHAPTER 8 Vectors

Before the Test

Review the ideas you listed in the table at the beginning of the chapter. Cross out any incorrect information in the first column. Then complete the table by filling in the third column.

K What I know...	W What I want to find out...	L What I Learned...

Math Online Visit *glencoe.com* to access your textbook, more examples, self-check quizzes, personal tutors, and practice tests to help you study for concepts in Chapter 8.

Are You Ready for the Chapter Test?

Use this checklist to help you study.

- ☐ I completed the Chapter 8 Study Guide and Review in the textbook.
- ☐ I took the Chapter 8 Practice Test in the textbook.
- ☐ I used the online resources for additional review options.
- ☐ I reviewed my homework assignments and made corrections to incorrect answers.
- ☐ I reviewed all vocabulary terms from the chapter.

Study Tip

To answer multiple-choice questions, read all of the answer choices first. Then cross out any choices that you know are not correct, and look for hints in other parts of the test for clues to the answer.

NAME ______________________ DATE ____________ PERIOD ____________

CHAPTER 9

Polar Coordinates and Complex Numbers

Before You Read

Before you read the chapter, respond to these statements.

1. Write an **A** if you agree with the statement.
2. Write a **D** if you disagree with the statement.

Before You Read	Polar Coordinates and Complex Numbers
	• The location of a point P in the polar coordinate system can be identified by polar coordinates of the form (r, θ).
	• The graph of a polar equation is either a circle or a line.
	• The equations $x = r \sin \theta$ and $y = r \cos \theta$ can be used to convert polar coordinates to rectangular coordinates.
	• A conic section with eccentricity $e > 0$, $d > 0$, and focus at the pole has a polar equation of the form $r = \frac{ed}{1 + e \cos \theta}$, $r = \frac{ed}{1 - e \cos \theta}$, $r = \frac{ed}{1 + e \sin \theta}$, or $r = \frac{ed}{1 - e \sin \theta}$.
	• De Moivre's Theorem states that $[r(\cos \theta + \boldsymbol{i} \sin \theta)]^n = rn(\cos \theta^n + \boldsymbol{i} \sin \theta^n)$.

Note-Taking Tips

- **Include an example of each type of polar graph in your notes. Indicate how the various dimensions of the figure are obtained.**
- **Review the new vocabulary in your notes every day.**

NAME ______________________ DATE __________ PERIOD __________

CHAPTER 9 Polar Coordinates and Complex Numbers

Key Points

Scan the pages in the chapter. Write at least one specific fact concerning each lesson. For example, in the lesson on polar coordinates, one fact might be that if r is positive, then P lies on the terminal side of θ. But if r is negative, P lies on the ray opposite the terminal side of θ. After completing the chapter, you can use this table to review for your chapter test.

Lesson	Fact
9-1 Polar Coordinates	
9-2 Graphs of Polar Equations	
9-3 Polar and Rectangular Forms of Equations	
9-4 Polar Forms of Conic Sections	
9-5 Complex Numbers and De Moivre's Theorem	

NAME ______________________ DATE __________ PERIOD __________

9-1 Polar Coordinates

What You'll Learn

Scan Lesson 9-1. List two headings that you would use to make an outline of this lesson.

1. ______________________________

____

2. ______________________________

____

Active Vocabulary

New Vocabulary Match each term with its definition by drawing a line to connect the two.

polar coordinate system	the origin in a polar coordinate system
polar axis	an equation expressed in terms of polar coordinates
polar coordinates	a two-dimensional coordinate system in which a point is located using a distance and an angle measure
polar equation	the set of all points with coordinates (r, θ) that satisfy a given polar equation
polar graph	coordinates of the form (r, θ), where r is the directed distance from the pole to the point and θ is the directed angle from the polar axis to $\overrightarrow{OP}$
pole	an initial ray from the pole that is usually horizontal and directed toward the right

Lesson 9-1

Main Idea	Details

Graph Polar Coordinates
pp. 534–536

Graph each point on the polar coordinate grid.

1. $P\left(2, \frac{2\pi}{3}\right)$

2. $Q\left(-1, \frac{3\pi}{4}\right)$

3. $R(3, 240°)$

4. $S\left(4, -\frac{3\pi}{2}\right)$

5. $T(-2, 225°)$

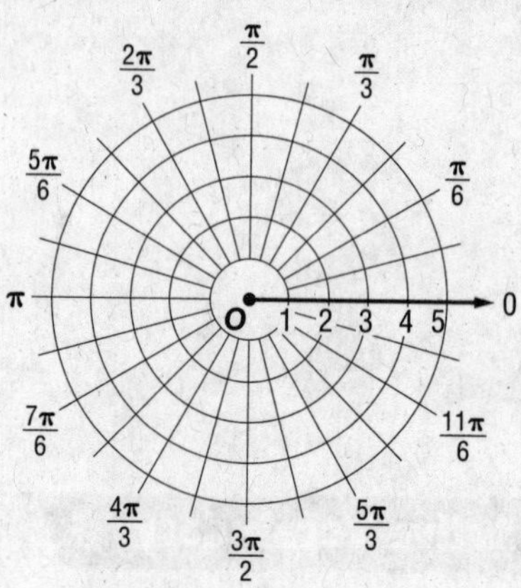

Graphs of Polar Equations
pp. 536–537

Graph each polar equation.

6. $r = 3$

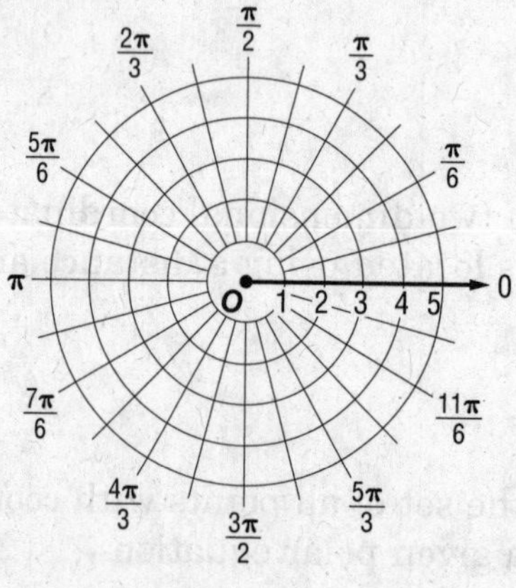

7. $\theta = \frac{3\pi}{4}$

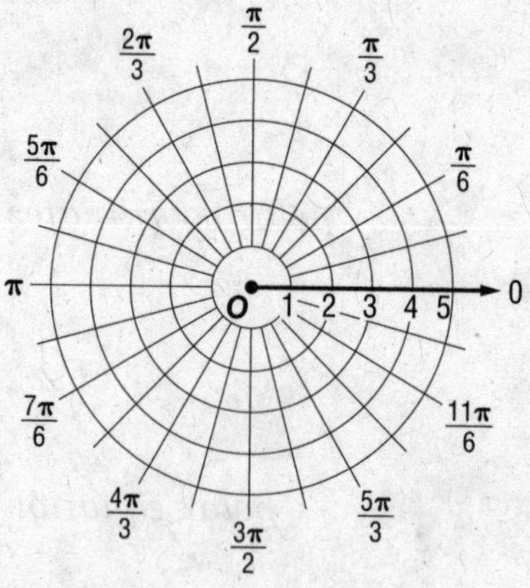

Find the distance between each pair of points.

1. $(2, 45°), (5, 105°)$ ______________

2. $(6, 220°), (-1, 100°)$ ______________

3. $\left(-6, \frac{2\pi}{3}\right), \left(-9, \frac{11\pi}{6}\right)$ ______________

4. $\left(3, \frac{\pi}{4}\right), \left(4, \frac{11\pi}{4}\right)$ ______________

NAME ______________________ DATE ____________ PERIOD ____________

9-2 Graphs of Polar Equations

What You'll Learn

Scan Lesson 9-2. Predict two things that you expect to learn based on the headings and Key Concept boxes.

1. ____

____

____

2. ____

____

____

Lesson 9-2

Active Vocabulary

New Vocabulary Fill in the blank with the correct term.

cardioid A curve that has an inner loop, comes to a point, has a dimple, or just curves outward is called a ______________.

lemniscate A curve that has a heart shape is called a ______________.

limaçon A curve that resembles a figure eight is called a ______________.

rose A curve that spirals out farther and farther from the pole at a constant rate is called a ______________.

spiral of Archimedes A curve that has three or more equal loops is called a ______________.

cardioid ▶ *(Lesson 9-2)* **Vocabulary Link** *Cardioid* begins with *cardio*. Look in a dictionary to find another word that begins with cardio. Explain how the English definition of this word can help you remember the meaning of cardioid.

____

____

____

Main Idea	Details

Graphs of Polar Equations
pp. 542–545

Graph each equation.

1. $r = 2 \sin \theta$ **2.** $r = 3 + 3 \cos \theta$ **3.** $r = 4 \sin 2\theta$

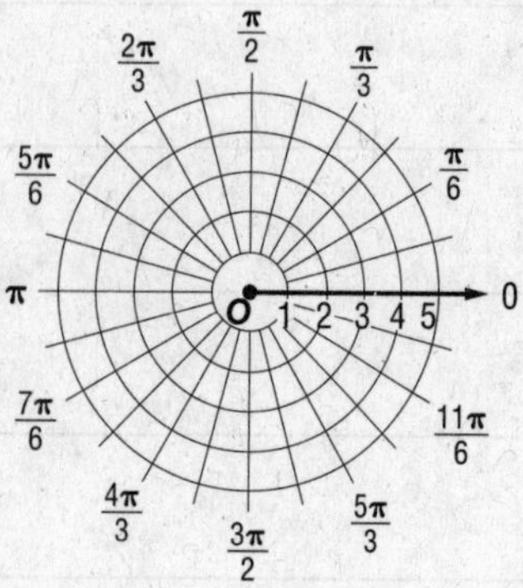
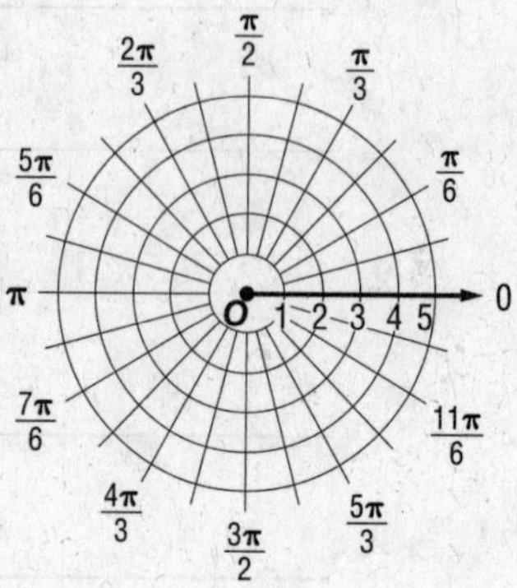
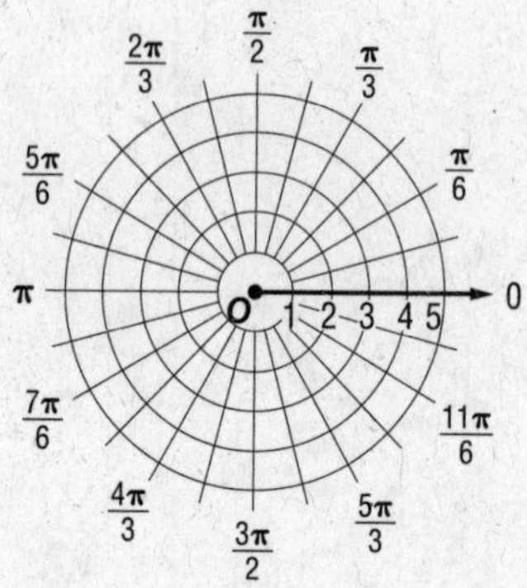

Classic Polar Curves
pp. 546–547

Write a polar equation, and sketch the graph of each classic curve.

rose with odd number of loops	spiral of Archimedes	lemniscate
______________________	______________________	______________________
______________________	______________________	______________________
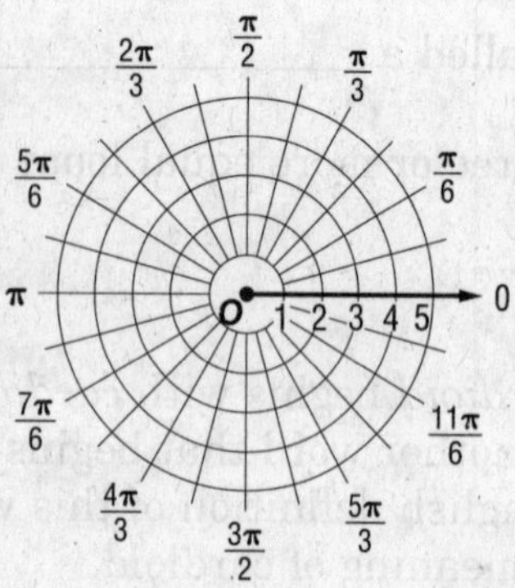	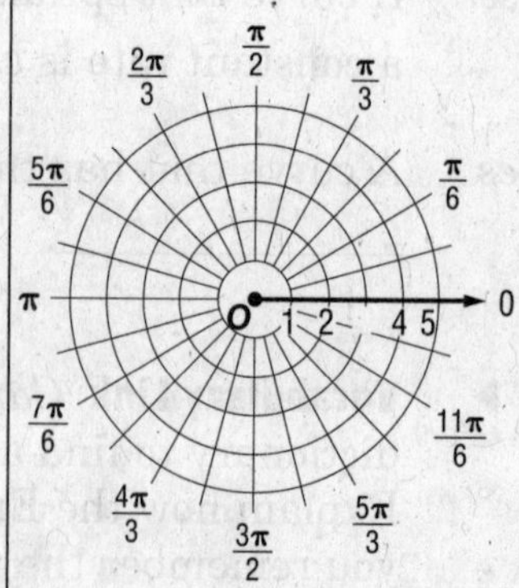	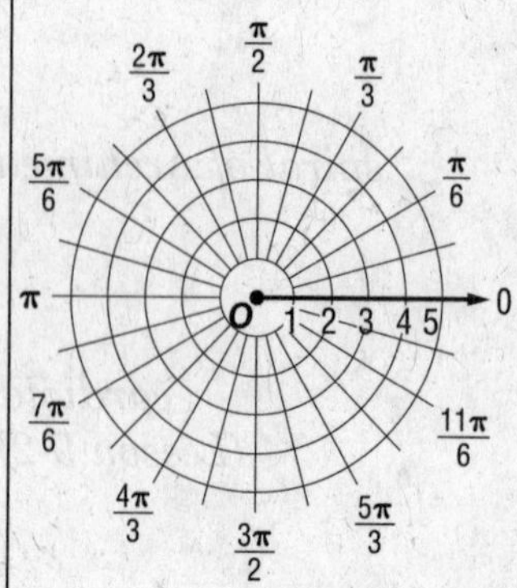

NAME ______________________ DATE __________ PERIOD __________

9-3 Polar and Rectangular Forms of Equations

What You'll Learn

Scan the examples for Lesson 9-3. Predict two things that you think you will learn about polar and rectangular forms of equations.

1. ______________________________

____

2. ______________________________

____

Active Vocabulary

Review Vocabulary Define each term in your own words.

polar coordinates ▶ (Lesson 9-1) ______________________________

____

____

____

polar equation ▶ (Lesson 9-1) ______________________________

____

polar axis ▶ (Lesson 9-1) ______________________________

____

Complete each trigonometric identity.

$\tan \theta =$ __________ $\cot \theta =$ __________

$\sec \theta =$ __________ $\csc \theta =$ __________

$\frac{1}{\cot \theta} =$ __________ $\cos^2 \theta +$ __________ $=$ __________

Lesson 9-3

Main Idea	Details

Polar and Rectangular Coordinates
pp. 551–554

Find the rectangular coordinates for each point with the given polar coordinates.

1. $P\left(2, \frac{\pi}{3}\right)$ ______________ **2.** $R(6, -30°)$ ______________

Find two pairs of polar coordinates for each point with the given rectangular coordinates.

3. $S(4\sqrt{2}, -4\sqrt{2})$ **4.** $T(-2\sqrt{3}, -2)$

______________ ______________

Polar and Rectangular Equations
pp. 554–556

Identify the graph of $y = (x - 2)^2 - 4$. Then write the equation in polar form.

____

____

Write $r = 6 \cos \theta$ in rectangular form and then identify its graph.

____

Helping You Remember

Explain how to convert a point given in rectangular coordinates to polar coordinates by graphing. Show an example.

____

____

____

____

____

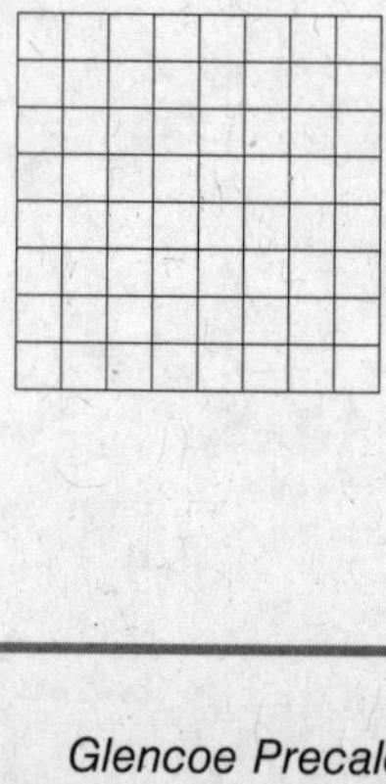

NAME ______________________ DATE ____________ PERIOD ____________

9-4 Polar Forms of Conic Sections

What You'll Learn

Scan the text in Lesson 9-4. Write two facts that you learned about polar forms of conic sections.

1. ____

____

____

2. ____

____

____

Active Vocabulary

Review Vocabulary Define each term in your own words.

eccentricity ▶ (Lesson 7-2) ____

____

____

ellipse ▶ (Lesson 7-2) ____

____

____

hyperbola ▶ (Lesson 7-3) ____

____

____

parabola ▶ (Lesson 7-1) ____

____

Lesson 9-4

Main Idea	Details

Use Polar Equations of Conics
pp. 561–563

Determine the eccentricity, type of conic, and equation of the directrix for each polar equation.

1. $r = \dfrac{15}{5 + 10 \cos \theta}$ ______________

2. $r = \dfrac{1}{4 - 2 \sin \theta}$ ______________

3. $r = \dfrac{6}{3 + 3 \cos \theta}$ ______________

Write Polar Equations of Conics
pp. 564–565

Write and graph a polar equation and directrix for the conic with $e = 3$ and directrix $x = 2$. Fill in the blanks.

Because $e = 3$, the conic is a ______________. The directrix, $x = 2$, is 2 to the ______________ of the pole, so the equation is of the form $r =$ ______________.

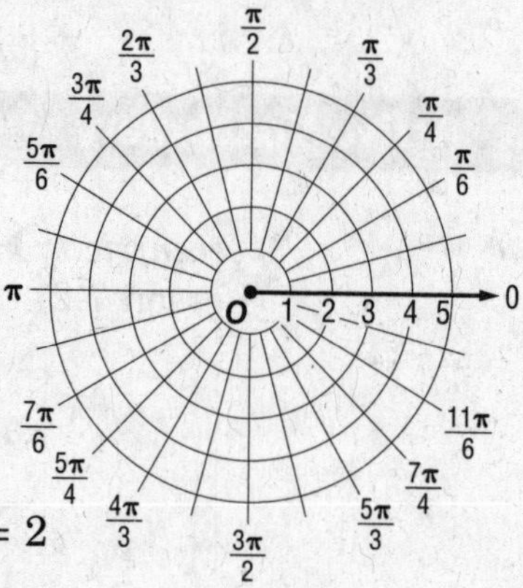

$r =$ ______________ $e = 3$ and $d = 2$

$=$ ______________ Simplify.

Helping You Remember

How can you remember that $r = \dfrac{ed}{1 \pm e \cos \theta}$ is oriented horizontally and $r = \dfrac{ed}{1 \pm e \sin \theta}$ is oriented vertically?

____

____

____

____

NAME ______________________ DATE __________ PERIOD __________

9-5 Complex Numbers and De Moivre's Theorem

What You'll Learn

Scan the lesson. Write two things that you already know about complex numbers.

1. ______________________________

2. ______________________________

Active Vocabulary

New Vocabulary Write the definition next to each term.

absolute value of a complex number ▶ ______________________________

argument ▶ ______________________________

complex plane ▶ ______________________________

imaginary axis ▶ ______________________________

polar form ▶ ______________________________

real axis ▶ ______________________________

Main Idea	Details

Polar Forms of Complex Numbers
pp. 569–571

Graph each number in the complex plane, and find its absolute value.

1. $z = -2 + 3i$ ________

2. $z = 7i$ ________

3. $z = 5 + i$ ________

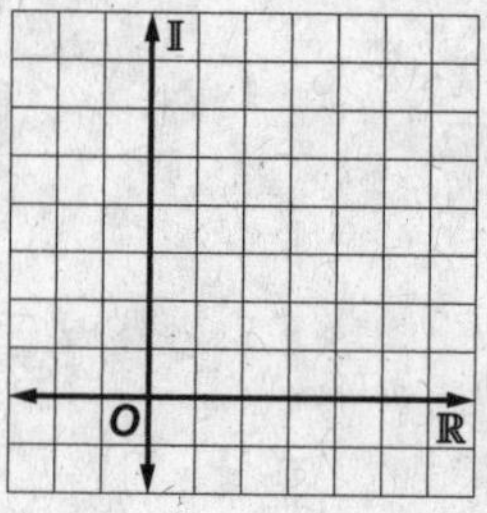

Products, Quotients, Powers, and Roots of Complex Numbers
pp. 572–577

Find $\dfrac{10\left(\cos \frac{5\pi}{6} + i \sin \frac{5\pi}{6}\right)}{2\left(\cos \frac{2\pi}{3} + i \sin \frac{2\pi}{3}\right)}$ **in polar form. Then express the quotient in rectangular form.**

$\dfrac{10\left(\cos \frac{5\pi}{6} + i \sin \frac{5\pi}{6}\right)}{2\left(\cos \frac{2\pi}{3} + i \sin \frac{2\pi}{3}\right)}$ Original expression

= ______________________ Quotient Formula

= ______________________ Simplify to polar form.

= ______________________ Evaluate and simplify.

Helping You Remember

How are the product and quotient formulas related to multiplication and division of monomial terms with exponents?

____

____

____

____

NAME ______________________ DATE __________ PERIOD __________

CHAPTER 9 Polar Coordinates and Complex Numbers

Tie It Together

Complete the details in the graphic organizers.

Polar Coordinates	Converting: Polar Equations and Rectangular Equations

Polar Graphs

Type	Equation	Graph
rose		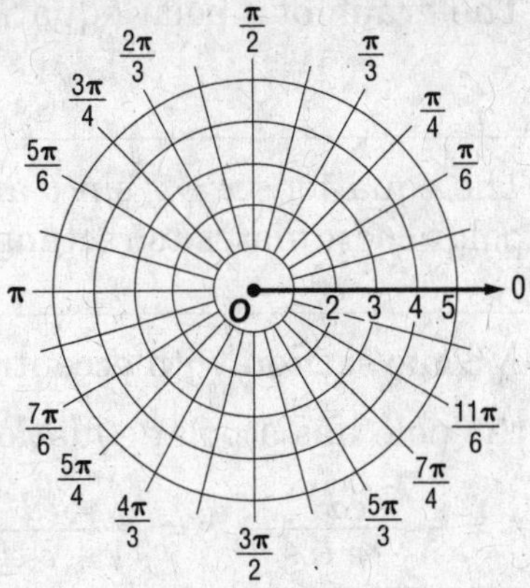
lemniscates		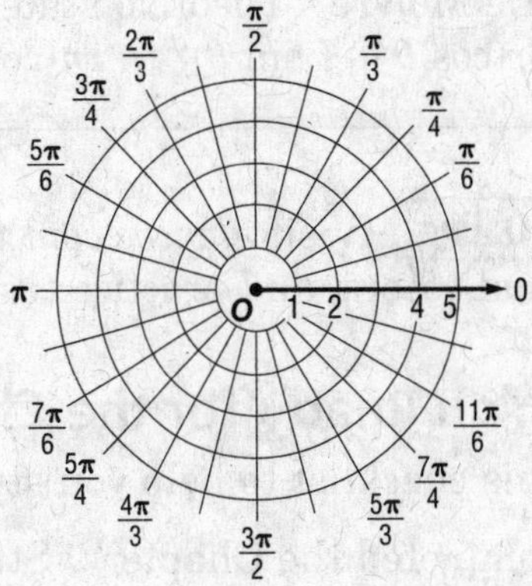

Complex Numbers

Product Formula:	DeMoivre's Theorem:
Quotient Formula:	Distinct Roots Formula:

NAME ______________________ DATE __________ PERIOD __________

CHAPTER 9

Polar Coordinates and Complex Numbers

Before the Test

Now that you have read and worked through the chapter, think about what you have learned. Complete the table below. Compare your previous answers with these.

1. Write an **A** if you agree with the statement.
2. Write a **D** if you disagree with the statement.

Polar Coordinates and Complex Numbers	After You Read
• The location of a point P in the polar coordinate system can be identified by polar coordinates of the form (r, θ).	
• The graph of a polar equation is either a circle or a line.	
• The equations $x = r \sin \theta$ and $y = r \cos \theta$ can be used to convert polar coordinates to rectangular coordinates.	
• A conic section with eccentricity $e > 0$, $d > 0$, and focus at the pole has a polar equation of the form $r = \frac{ed}{1 + e \cos \theta}$, $r = \frac{ed}{1 - e \cos \theta}$, $r = \frac{ed}{1 + e \sin \theta}$ or $r = \frac{ed}{1 - e \sin \theta}$.	
• De Moivre's Theorem states that $[r(\cos \theta + \boldsymbol{i} \sin \theta)]^n = rn(\cos \theta^n + \boldsymbol{i} \sin \theta^n)$.	

Math Online Visit *glencoe.com* to access your textbook, more examples, self-check quizzes, personal tutors, and practice tests to help you study for concepts in Chapter 9.

Are You Ready for the Chapter Test?

Use this checklist to help you study.

☐ I completed the Chapter 9 Study Guide and Review in the textbook.

☐ I took the Chapter 9 Practice Test in the textbook.

☐ I used the online resources for additional review options.

☐ I reviewed my homework assignments and made corrections to incorrect answers.

☐ I reviewed all vocabulary terms from the chapter.

Study Tip

- When studying for tests, create and use graphic organizers to show relationships between concepts.

CHAPTER 10 Sequences and Series

Before You Read

Before you read the chapter, respond to the following statements.

1. Write an **A** if you agree with the statement.

2. Write a **D** if you disagree with the statement.

Before You Read	Sequences and Series
	• A sequence is a function with a domain that is the set of natural numbers.
	• An arithmetic series is the sum of the terms of a geometric sequence.
	• To find the common ratio for a geometric sequence, divide any term by the previous term.
	• When using the principle of mathematical induction to prove a conjecture, the anchor step is also known as the inductive hypothesis.
	• The Binomial Theorem can be used to expand a binomial.

Note-Taking Tips

- **Remember to study your notes daily.**
  Reviewing small amounts at a time will help you retain the information.
- **It is helpful to read through your notes before beginning your homework.**
  Look over any page-referenced material.

NAME ______________________________ DATE ____________ PERIOD ____________

CHAPTER 10 Sequences and Series

Key Points

Scan the pages in the chapter. Write at least one specific fact concerning each lesson. For example, in the lesson on arithmetic sequences and series, one fact might be that in an arithmetic sequence the difference between successive terms is constant. After completing the chapter, you can use this table to review for your chapter test.

Lesson	Fact
10-1 Sequences, Series, and Sigma Notation	
10-2 Arithmetic Sequences and Series	
10-3 Geometric Sequences and Series	
10-4 Mathematical Induction	
10-5 The Binomial Theorem	
10-6 Functions as Infinite Series	

NAME ______________________ DATE __________ PERIOD __________

10-1 Sequences, Series, and Sigma Notation

What You'll Learn

Scan the lesson. Write two things that you already know about sequences and series.

1. ____

____

2. ____

____

Active Vocabulary

New Vocabulary Label the diagram with a term listed at the left.

explicit sequence

finite sequence

finite series

infinite sequence

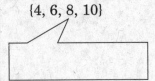

infinite series

recursive sequence

sigma notation

$\{4, 6, 8, 10\}$ [ ]

$-4 + (-1) + 2 + 5 + \ldots$ [ ]

$a_n = a_{n-1} + 7, a_1 = -3$ [ ]

$\sum_{n=1}^{4} 2n + 2$ [ ]

$4 + 6 + 8 + 10$ [ ]

$a_n = 5n - 1$ [ ]

[ ] $\{-4, -1, 2, 5, \ldots\}$

Main Idea	Details

Sequences
pp. 590–592

Write each sequence in the appropriate box. Then find the sixth term of each sequence.

$a_n = 3a_{n-1} + 2, a_1 = -5$ $\quad b_n = n^2 + n$

$c_n = \dfrac{n+1}{2}$ $\quad d_n = 2d_{n-2} - d_{n-1}, d_1 = 2, d_2 = 6$

Explicit	Recursive

Series and Sigma Notation
pp. 593–594

Find the fourth partial sum of $a_n = \dfrac{3}{4^n}$.

Find the first four terms.

$a_1 =$ ___ or ___ $\quad a_3 =$ ___ or ___

$a_2 =$ ___ or ___ $\quad a_4 =$ ___ or ___

The fourth partial sum is $S_4 =$ ______________ or ______

Helping You Remember

In your own words, explain the difference between a sequence and a series. Provide an example of each.

____

____

____

Sequence: $\qquad$ Series:

NAME ______________________ DATE __________ PERIOD __________

10-2 Arithmetic Sequences and Series

What You'll Learn

Scan the examples for Lesson 10-2. Predict two things that you think you will learn about arithmetic sequences and series.

1. ____

____

____

2. ____

____

____

Lesson 10-2

Active Vocabulary

New Vocabulary Match each term with its definition by drawing a line to connect the two.

arithmetic means	the difference found by subtracting any term in an arithmetic sequence from its succeeding term
arithmetic sequences	the common difference in an arithmetic sequence
arithmetic series	the terms between two nonconsecutive terms of an arithmetic sequence
common difference	a set of numbers in which the difference between successive terms is a constant
first difference	the sum of the terms of an arithmetic sequence
second difference	the differences of consecutive first differences

Main Idea	Details

Arithmetic Sequences
pp. 599–602

Write an arithmetic sequence that has five arithmetic means between 7 and −2. Note that $a_7 = -2$.

First, find the common difference using $a_7 =$ ____, $a_1 =$ ____, and $n =$ ____.	Then determine the arithmetic means by using $d =$ ____.
$a_n = a_1 + (n - 1)d$	$a_2 = 7 +$ ____ $=$ ____
____ $=$ ____ $+ ($____ $- 1)d$	$a_3 =$ ____ $+ -1.5 =$ ____
$-2 = 7 + 6d$	$a_4 = 4 +$ ____ $= 2.5$
$d =$ ____	$a_5 =$ ____ $+ -1.5 = 1$
	$a_6 =$ ____ $+$ ____ $=$ ____

The sequence is 7, ____, ____, ____, ____, ____, −2.

Arithmetic Series
pp. 602–604

Find $\sum_{n=8}^{32} 4n - 11$.

Find a_1, a_n, and n.

$n =$ ______________ Upper bound minus lower bound plus 1

$a_1 =$ ______________ Lower bound = ______________

$=$ ______________ Simplify.

$a_{25} =$ ______________ Upper bound = ______________

$=$ ______________ Simplify.

Use the first sum of a finite arithmetic series formula.

$S_n =$ ______________ Sum of a finite arithmetic series formula

$S_{25} =$ ______________ $n = 25$, $a_1 = 21$, and $a_n = 117$

$S_{25} =$ ______________ Simplify.

Therefore, ______________

10-3 Geometric Sequences and Series

What You'll Learn

Scan the text in Lesson 10-3. Write two facts that you learned about geometric sequences and series.

1. ____

____

____

2. ____

____

____

Active Vocabulary

Review Vocabulary Define *sequence* in your own words. (Lesson 10-1)

____

____

____

New Vocabulary Fill in each blank with the correct term.

common ratio — A ______________________ is the sum of the terms of a geometric sequence.

geometric means — A sequence in which the ratio between successive terms is a constant is called a ______________________.

geometric sequence — The ______________________ in a geometric sequence is the constant found by dividing any term in the sequence by the preceding term.

geometric series — The terms between nonconsecutive terms of a geometric sequence are known as ______________________.

Lesson 10-3

Main Idea	Details

Geometric Sequences
pp. 608–611

Use the geometric sequence 100, 10, 1, 0.1, ... to complete the organizer.

Step 1: Identify the first term of the sequence.

____

↓

Step 2: Find the common ratio.

____

↓

Step 3: Write an explicit formula for finding the nth term of the sequence.

____

↓

Step 4: Write a recursive formula for finding the nth term of the sequence.

____

____

↓

Step 5: Find the next three terms in the sequence.

____

Geometric Series
pp. 611–614

For each infinite geometric series, find the common ratio. Then write *yes* or *no* to indicate whether each series has a sum. If the series has a sum, find it.

Series	Common Ratio	Sum? *yes* or *no*	Sum
40, 20, 10, 5, ...			
−3, −6, −12, −24, ...			
$\frac{1}{3}, \frac{1}{9}, \frac{1}{27}, \frac{1}{81}, \ldots$			

NAME ______________________ DATE ____________ PERIOD ____________

10-4 Mathematical Induction

What You'll Learn

Scan the text under the *Now* heading. List two things that you will learn in the lesson.

1. ______________________________

2. ______________________________

Active Vocabulary

New Vocabulary Write the definition next to each term.

anchor step ▶ ______________________________

extended principle of mathematical induction ▶ ______________________________

inductive hypothesis ▶ ______________________________

inductive step ▶ ______________________________

principle of mathematical induction ▶ ______________________________

Lesson 10-4

Main Idea	Details

Mathematical Induction
pp. 621–623

Prove that $2^{3n} - 4$ is divisible by 4 for all positive integers n.

Conjecture	→	Let P_n be the statement that $2^{3n} - 4$ is divisible by 4 for all positive integers n.
Anchor step Verify that P_n is true for $n = 1$.	→	
	→	Assume that $2^{3k} - 4 = 4r$ is true for some integer r.
Inductive Step Use the inductive hypothesis to prove that P_n is true for $n = k + 1$.	→	
Conclusion	→	

Extended Mathematical Induction
p. 624

Identify which conjectures require the use of the extended principle of mathematical induction to prove. Write *yes* or *no* on the answer space provided.

1. $3n - 4 \geq n$, for all $n \geq 2$ __________

2. $2 + 4 + 6 + \ldots + 2n = n^2 + n$ __________

3. $2^n \geq n^2$, for all $n \geq 4$ __________

NAME ______________________ DATE ____________ PERIOD ____________

10-5 The Binomial Theorem

What You'll Learn

Scan Lesson 10-5. Predict two things that you expect to learn based on the headings and Key Concept boxes.

1. ______________________

2. ______________________

Active Vocabulary

Review Vocabulary Define *monomial function* in your own words. (Lesson 2-1)

____

Define *polynomial function* in your own words. (Lesson 2-2)

____

New Vocabulary Write the correct term next to each definition.

____________ ▶ a formula used to expand powers of binomials

____________ ▶ a pattern of numbers that displays the coefficients of the expansion of $(a + b)^n$, where n is an element of the set of whole numbers

____________ ▶ constant numbers that are multiplied by the terms in a binomial expansion

Main Idea	Details

Pascal's Triangle
pp. 628–629

Use Pascal's triangle to expand $(a + b)^5$.

Step 1: Write the series for $(a + b)^5$, omitting the coefficients.

____

Step 2: Identify the numbers in the 5th row of Pascal's triangle.

____

Step 3: Use the numbers from the 5th row of Pascal's triangle as the coefficients of the terms.

$(a + b)^5 =$ ______________________________

Binomial Theorem
pp. 630–632

Complete the equations by using the Binomial Theorem to expand $(2x - y)^3$.

Expand $(a + b)^3$.

$(a + b)^3 =$ ____ $a^3b^0 +$ ____ $a^2b^1 +$ ____ $a^1b^2 +$ ____ a^0b^3

$=$ ____ $a^3 +$ ____ $a^2b +$ ____ $ab^2 +$ ____ b^3

Substitute $a = 2x$ and $b = -y$.

$=$ ______________________________

$=$ ________ $+$ ________ $+$ ________ $+$ ________

Helping You Remember

When finding the binomial coefficients using the Binomial Theorem, you must evaluate ${}_nC_r$ for various values of n and r. What does ${}_nC_r$ equal? Choose a particular value for n and r and illustrate.

____

____

____

10-6 Functions as Infinite Series

What You'll Learn

Scan Lesson 10-6. List two headings that you would use to make an outline of this lesson.

1. ______________________________

2. ______________________________

____

Active Vocabulary

Review Vocabulary Define *series* in your own words. (Lesson 10-1)

____

____

Define *infinite series* in your own words. (Lesson 10-1)

____

____

New Vocabulary Fill in each blank with the correct term.

Euler's Formula — An infinite series of the form $\sum_{n=0}^{\infty} a_n x^n = a_0 + a_1 x + a_2 x^2 + a_3 x^3 + \dots$, where x and a can take on any values and $n = 0, 1, 2, \dots$, is called a(n) ______________ in x.

exponential series — The power series representing $\sin x$ and $\cos x$ are called ______________.

power series — The power series representing e^x is called the ______________.

trigonometric series — For any real number θ, the relationship $e^{i\theta} = \cos\theta + \boldsymbol{i}\sin\theta$ is known as ______________.

Main Idea | **Details**

Power Series
pp. 636–637

Complete each box by using $\sum_{n=0}^{\infty} x^n$ to find a power series representation of $g(x) = \frac{1}{4-x}$. Indicate the interval on which the series converges.

What you know:

$f(x) = \frac{1}{1-x} =$ ______________

$g(x) = \frac{1}{4-x}$

What you need to find out:

$g(x) = f(u)$

$\frac{1}{4-x} = \frac{1}{1-u}$

$1 - u =$ ______________

$u =$ ______________

Find $g(x)$ in terms of $f(x)$:

$g(x) =$ ______________

Finally represent $g(x)$ as a power function:

$g(x) = \frac{1}{4-x}$

$=$ ______________;

____

Transcendental Functions as Power Series
pp. 638–641

Use the fifth partial sum of the trigonometric series for cosine to approximate the value of $\cos \frac{\pi}{5}$ to three decimal places.

$\cos x =$ ______________ $=$ ______________

$\cos \frac{\pi}{5} \approx$ ______________

$\approx$ ______________ Simplify.

Use a calculator to check your answer. The approximation is ______________ because $\cos \frac{\pi}{5} \approx$ ______________.

NAME ________________ DATE ________ PERIOD ________

CHAPTER 10 Sequences and Series

Tie It Together

Provide an example for each situation that can be used to complete the table.

<table>
<tr><th colspan="2">Finite Arithmetic Sequence</th><th colspan="2">Infinite Geometric Sequence</th></tr>
<tr><td colspan="2"></td><td colspan="2"></td></tr>
<tr><td colspan="2">Identify the common difference:</td><td colspan="2">Identify the common ratio:</td></tr>
<tr><td>Explicit form for finding the nth term:</td><td>Recursive form for finding the nth term:</td><td>Explicit Form for Finding the nth Term:</td><td>Recursive Form for Finding the nth Term:</td></tr>
<tr><td colspan="2">Write the sequence as a series:</td><td colspan="2">Write the sequence as a series.</td></tr>
<tr><td rowspan="2">Find the sum given a_1 and a_n:</td><td rowspan="2">Find the sum given a_1 and d:</td><td colspan="2">Does your series have a sum?
If so, why?</td></tr>
<tr><td colspan="2">Find the sum, if it exists.</td></tr>
</table>

Rewrite Pascal's triangle, shown to the left, using the notation ${}_nC_r$.

			1				0th row	____
		1		1			1st row	____ ____
	1		2		1		2nd row	____ ____ ____
1		3		3		1	3rd row	____ ____ ____ ____

NAME ______________________________ DATE ____________ PERIOD ____________

CHAPTER 10 Sequences and Series

Before the Test

Now that you have read and worked through the chapter, think about what you have learned and complete the table below. Compare your previous answers with these.

1. Write an **A** if you agree with the statement.
2. Write a **D** if you disagree with the statement.

Sequences and Series	After You Read
• A sequence is a function with a domain that is the set of natural numbers.	
• An arithmetic series is the sum of the terms of a geometric sequence.	
• To find the common ratio for a geometric sequence, divide any term by the previous term.	
• When using the principle of mathematical induction to prove a conjecture, the anchor step is also known as the inductive hypothesis.	
• The Binomial Theorem can be used to expand a binomial.	

Math Online Visit *glencoe.com* to access your textbook, more examples, self-check quizzes, personal tutors, and practice tests to help you study for concepts in Chapter 10.

Are You Ready for the Chapter Test?

Use this checklist to help you study.

☐ I completed the Chapter 10 Study Guide and Review in the textbook.

☐ I took the Chapter 10 Practice Test in the textbook.

☐ I used the online resources for additional review options.

☐ I reviewed my homework assignments and made corrections to incorrect answers.

☐ I reviewed all of the vocabulary terms from the chapter.

Study Tip

- Create a mnemonic device to remember lists of sequences. For example, **P**lease **E**xcuse **M**y **D**ear **A**unt **S**ally is an acronym for remembering the order of operations: **p**arentheses, **e**xponents, **m**ultiply and **d**ivide, **a**dd and **s**ubtract.

NAME ____________________ DATE ____________ PERIOD ____________

CHAPTER 11 Inferential Statistics

Before You Read

Before you read the chapter, respond to these statements.

1. Write an **A** if you agree with the statement.

2. Write a **D** if you disagree with the statement.

Before You Read	Inferential Statistics
	• The median of a set of data is less affected by the presence of outliers than the mean.
	• There are three types of random variables: discrete, continuous, and infinite.
	• The total area under a normal distribution curve is 10.
	• The degrees of freedom represent the number of values that are fixed after a sample statistic is determined.
	• The correlation coefficient determines the type and strength of the linear relationship between two variables.

Note-Taking Tips

- **When taking notes, summarize the main ideas presented in the lesson.** Summaries are helpful for realizing what is important.
- **Copy your notes after class while the concepts are still fresh in your mind.**

NAME ______________________ DATE ____________ PERIOD ____________

CHAPTER 11 Inferential Statistics

Key Points

Scan the pages in the chapter. Write at least one specific fact concerning each lesson. For example, in the lesson on descriptive statistics, one fact might be that in skewed distributions the mean is located closer to the tail than the median. After completing the chapter, you can use this table to review for your chapter test.

Lesson	Fact
11-1 Descriptive Statistics	
11-2 Probability Distributions	
11-3 The Normal Distribution	
11-4 The Central Limit Theorem	
11-5 Confidence Intervals	
11-6 Hypothesis Testing	
11-7 Correlation and Linear Regression	

NAME ______________________ DATE __________ PERIOD __________

11-1 Descriptive Statistics

What You'll Learn

Scan the lesson. Write two things that you already know about statistics.

1. ______________________

____

2. ______________________

____

Active Vocabulary

New Vocabulary Label each diagram with a term listed at the left.

negatively skewed distribution

symmetrical distribution

positively skewed distribution

____

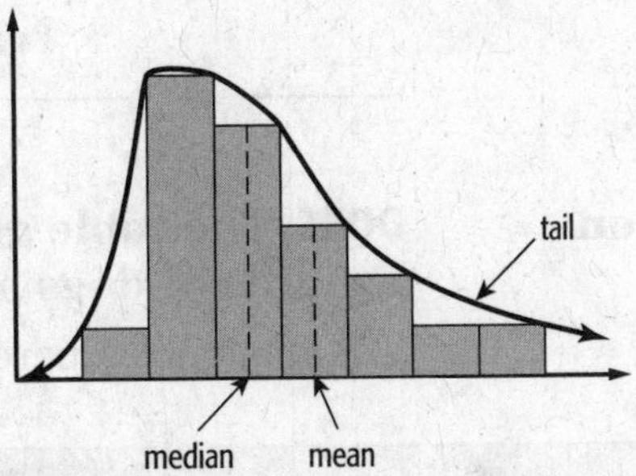

____

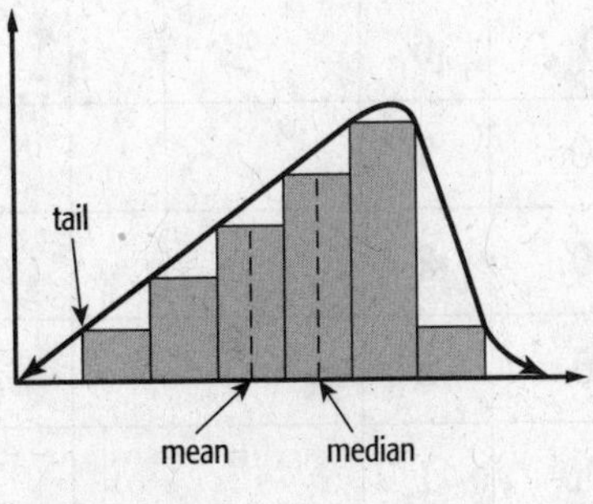

____

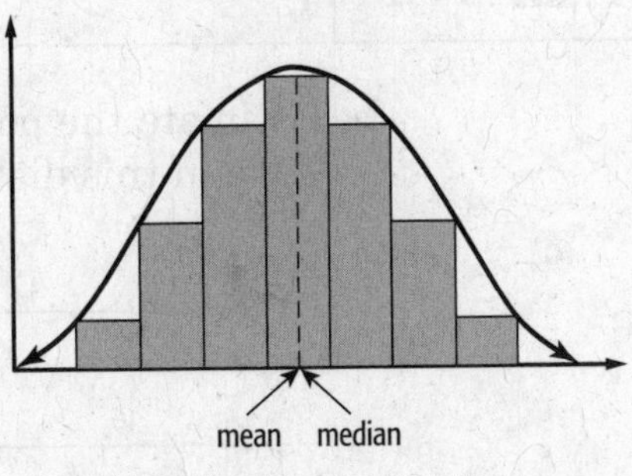

Lesson 11-1

Main Idea	Details

Describing Distributions
pp. 654–657

ELECTRONICS The table below shows the prices of televisions at an electronics store. Summarize the center and spread of the data using either the standard deviation or the five-number summary.

Television Prices ($)				
349	250	455	375	550
320	476	1980	2295	750
600	799	1200	964	1095

____

____

____

Measures of Position
pp. 658–659

DOGS The table gives the frequency distribution of the weights of dogs at a kennel.

a. Construct a percentile graph of the data.

Class Boundaries	*f*	Cumulative Frequency	Cumulative Percentages
1–20	6		
21–40	7		
41–60	3		
61–80	2		
81–100	1		
101–120	1		

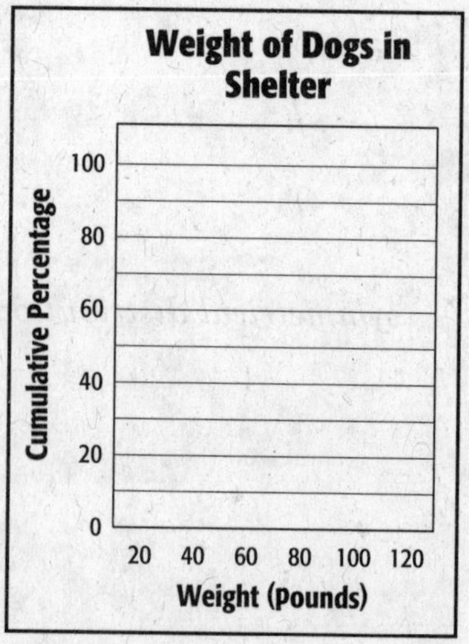

b. Estimate the percentile rank that a 52-pound dog would have in this distribution and interpret its meaning.

____

____

NAME ______________________ DATE ____________ PERIOD ____________

11-2 Probability Distributions

What You'll Learn

Scan the text in Lesson 11-2. Write two facts that you learned about probability distributions.

1. ____

____

____

____

2. ____

____

____

____

Lesson 11-2

Active Vocabulary

New Vocabulary Fill in each blank with the correct term.

binomial distribution — A(n) ______________ represents a numerical value assigned to an outcome of a probability experiment.

continuous random variable — A(n) ______________________ can take on an infinite number of possible values within a specified interval.

discrete random variable — A(n) ____________________ is a table, equation, or graph that links each possible value with its probability of occurring.

expected value — The ____________ of a random variable for a probability distribution is equal to the mean of the random variable.

probability distribution — The distribution of the outcomes of a binomial experiment is called a(n) ______________________.

random variable — A(n) ______________________ can take on a finite or countable number of possible values.

Main Idea | **Details**

Probability Distributions
pp. 664–667

SLEEP The table shows the number of nights that students got at least 8 hours of sleep last week. Construct a probability distribution for the random variable X. Then find and interpret the mean in the context of the given situation, and find the variance and standard deviation.

Nights, X	Frequency	$P(X)$	$X\,P(X)$	$(X-\mu)^2$	$(X-\mu)^2\,P(X)$
0	2				
1	5				
2	7				
3	12				
4	28				
5	32				
6	20				
7	14				

The mean is _____, which indicates that ____________

____

____

The variance is __________ and the standard deviation __________.

Binomial Distribution
pp. 668–670

Determine whether the following experiment is a binomial experiment or can be reduced to a binomial experiment. If it is, state the values of n, p, and q. If it is not, explain why not.

The results of an airline survey show that 20% of passengers carry laptop computers on a flight. Ten passengers are randomly selected and asked whether they are carrying laptop computers. The random variable represents those who say they have laptop computers.

____

____

____

NAME ______________________ DATE __________ PERIOD __________

11-3 The Normal Distribution

What You'll Learn

Scan Lesson 11-3. List two headings that you would use to make an outline of this lesson.

1. ______________________

2. ______________________

Active Vocabulary

Review Vocabulary Define each term in your own words.

continuous random variable (Lesson 11-2) ▶ ______________________

New Vocabulary Write the definition next to each term.

normal distribution ▶ ______________________

empirical rule ▶ ______________________

z-value ▶ ______________________

standard normal distribution ▶ ______________________

Lesson 11-3

Main Idea	Details
The Normal Distribution pp. 674–678	**Complete the diagram by filling in the percents of the data that fall in each area.**

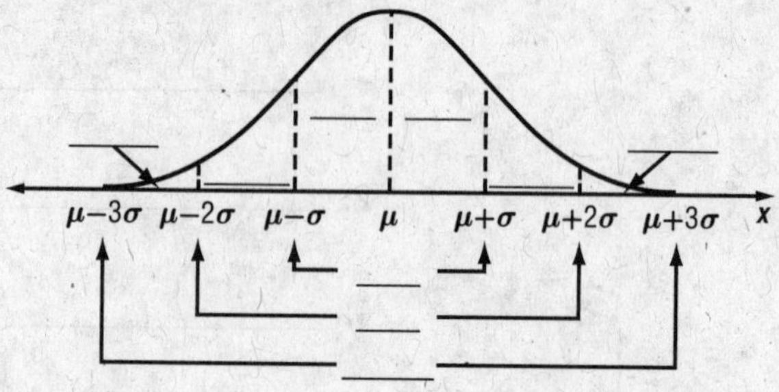

Find each value.

1. z if $X = 36$, $\mu = 38$, and $\sigma = 3.6$ ____________
2. X if $z = 1.56$, $\mu = 12$, and $\sigma = 2.4$ ____________

Find the interval of z-values associated with each area.

3. the middle 40% of the data ____________
4. the outside 80% of the data ____________

Probability and the Normal Distribution
pp. 679–680

METEOROLOGY The temperatures for one month for a city in Oregon are normally distributed with $\mu = 62°$ and $\sigma = 5°$. Find the probability and use a graphing calculator to sketch the corresponding area under the curve for $P(50° < X < 70°)$.

____________ Formula for z-values ____________

= ____________ Substitute. = ____________

= ____________ Simplify. = ____________

Select [2nd] [VARS]. Under the **DISTR** menu, select **normalcdf** (____________). The area between z = ____________ and z = ____________ is about ____________.

Therefore, approximately ____________ of the temperatures were between 50° and 70°.

NAME ______________________ DATE ____________ PERIOD ____________

11-4 The Central Limit Theorem

What You'll Learn

Scan the text under the *Now* heading. List two things that you will learn in the lesson.

1. ______________________

____

____

2. ______________________

____

____

Active Vocabulary

Review Vocabulary Define *z-value* in your own words. (Lesson 11-3)

____

____

____

New Vocabulary Write the correct term next to each definition.

______________ ▶ a correction for continuity that is used when approximating a binomial distribution

______________ ▶ differences between sample means and the population mean that occur because the sample is not a complete representation of the population

______________ ▶ a distribution of the means of random samples of a certain size that are taken from a population

______________ ▶ the standard deviation of the sample means

Lesson 11-4

Main Idea	Details
The Central Limit Theorem pp. 685–688	**MOVIES The average prices for movie tickets are \$6.50 for a child, \$8.50 for an adult, and \$6.00 for a senior. For a random sample of 40 theaters, find the probability that the mean ticket price is more than \$8.65 for an adult if the standard deviation is \$0.50.** ______________________________
The Normal Approximation pp. 689–691	**PROM A school survey reported that 32% of the senior class was planning to attend prom. If 40 seniors are randomly selected, find the probability that fewer than 8 will attend prom. Complete the following steps.**

In this binomial experiment, $n =$ ___, $p =$ ____, and $q =$ ____.

Step 1 ______ Mean and standard deviation formulas. ______

$=$ ________ Simplify. $=$ ________

Since np and nq ____________________, the normal distribution ____________________ approximate the binomial distribution.

Step 2 Write the problem in probability notation: ________

Step 3 Rewrite the problem with the continuity factor included:

____

Step 4 Find the corresponding z-value for $X =$ ________.

$z = \frac{\quad}{\quad}$ z-value formula

$= \frac{\quad}{\quad}$ or ________

Step 5 Use a graphing calculator to find the corresponding area between $z =$ ________ and $z =$ ________.

The approximate area that corresponds to the area below $z = -1.80$ is ________. Therefore, the probability that fewer than 8 seniors will be attending prom is about ________.

11-5 Confidence Intervals

What You'll Learn

Scan the Examples for Lesson 11-5. Predict two things that you will learn about confidence intervals.

1. ______________________________

____

____

2. ______________________________

____

____

Active Vocabulary

New Vocabulary Match the term with its definition by drawing a line to connect the two.

inferential statistics	*z*-values that correspond to a particular confidence level
confidence interval	a method where a sample of data is analyzed and conclusions are made about the entire population
critical values	the number of values that are free to vary after a sample statistic is determined
t-distribution	a family of curves that are dependent on a parameter known as the degrees of freedom
degrees of freedom	a measure that describes a characteristic of a population
parameter	a specific interval estimate of a parameter

Main Idea	Details

Normal Distribution
pp. 696–698

TEXTING A poll of 36 randomly selected high school students showed that they spend an average of 22 minutes a day texting. Assume a standard deviation of 7 minutes. Find a 95% confidence interval for the mean texting time for all of the students. Complete the following steps.

$CI =$ ______________ Confidence Interval for the Mean

$=$ ______________ $\overline{x} =$ ____________, $z =$ ____________,

$\sigma =$ ____________, $n =$ ____________

$\approx$ ______________ Simplify.

Left boundary: __________ Right boundary: __________

The 95% confidence interval is ______________________.

***t*-Distribution**
pp. 699–701

Use the *t*-distribution to find each confidence interval, given the following information.

1. 90%, $\overline{x} = 18.4$, $s = 0.9$, $n = 15$ ____________________

2. 95%, $\overline{x} = 72$, $s = 10.9$, $n = 28$ ____________________

3. 99%, $\overline{x} = 93.1$, $s = 4.3$, $n = 22$ ____________________

4. 95%, $\overline{x} = 15.2$, $s = 2.6$, $n = 19$ ____________________

Helping You Remember

Write four characteristics of the *t*-distribution.

____

____

____

____

NAME ______________________ DATE ____________ PERIOD ____________

11-6 Hypothesis Testing

What You'll Learn

Scan Lesson 11-6. Predict two things that you expect to learn based on the headings and Key Concept box.

1. ______________________________

____

____

2. ______________________________

____

____

Active Vocabulary

New Vocabulary Fill in the blank with the correct term.

hypothesis test The ____________ states that there is not a significant difference between a sample value and the population parameter.

null hypothesis The ____________ is the maximum allowable probability of committing a type 1 error.

alternative hypothesis If H_a: $\mu \neq k$, the hypothesis test is a(n) ____________.

level of significance A(n) ____________ assesses evidence provided by data about a claim concerning a population parameter.

two-tailed test The ____________ is the lowest level of significance at which H_0 can be rejected for a given set of data.

p-value The ____________ states that there is a difference between a sample value and the population parameter.

Lesson 11-6

Main Idea	Details

Hypotheses
p. 705

Write the null and alternative hypotheses for each statement, and state which hypothesis represents the claim.

1. Lee claims that he owns at least 100 pairs of shoes.

____

2. An airline claims that more than 90% of its flights arrive on schedule.

____

Significance and Tests
pp. 706–709

DVDS A company claims that its DVD player will run for 6 hours or more on a fully charged battery. A random sample of 50 DVD players shows a mean of 5.8 hours with a standard deviation of 0.6 hour. Is there enough evidence to reject the claim at $\alpha = 0.05$? Complete the following steps to solve the problem.

Step 1 State the null and alternative hypotheses and identify the claim.

H_0: ______________ H_a: ______________

Step 2 Determine the critical value(s) and region.

The population standard deviation is __________
and __________, so you can use the __________.
The test is __________ since __________.
Let α = __________. The z-value is __________.

Step 3 Calculate the test statistic.

____

Step 4 Accept or reject the null hypothesis.

H_0 is __________ since the test statistic does
__________ within the critical region. Therefore,
there is __________ to reject the claim.

NAME ______________________ DATE ____________ PERIOD ____________

11-7 Correlation and Linear Regression

What You'll Learn

Scan the text in Lesson 11-7. Write two facts that you learned about correlation and linear regression.

1. ______________________________

2. ______________________________

Active Vocabulary

New Vocabulary Write the definition next to each term.

correlation coefficient ▶ ______________________________

regression line ▶ ______________________________

residual ▶ ______________________________

interpolation ▶ ______________________________

extrapolation ▶ ______________________________

Main Idea	Details

Correlation
pp. 713–715

TELEVISON AND GPA Marty surveyed students in her school regarding their grade point averages and the amount of television they watch per day. The results are shown below. Make a scatter plot of the data and identify the relationship. Then calculate and interpret the correlation coefficient, and determine if the correlation coefficient is significant at the 5% level.

Hours watched	6.5	3.2	1.6	5.4	3.9	4.8	0.2	1.6	2.3	2.5
GPA	1.9	3.0	3.6	2.2	3.1	2.5	3.9	3.7	3.3	3.0
Hours watched	1.3	2.9	3.4	3.6	4.0	2.8	1.5	3.0	5.6	4.2
GPA	3.5	2.8	2.9	2.5	2.3	2.6	3.1	2.8	1.6	2.2

Correlation: ______________________

Correlation coefficient: ______________________

H_0: __________ H_a: __________

Critical values: $\alpha =$ __________ with __________ degrees of freedom and $t =$ __________

Calculate the test statistic.

$t =$ __________

Since ______________,

the null hypothesis is ________.

The equation of the least-squares regression line is

______________________.

Linear Regression
pp. 716–719

Use the regression equation to predict the expected GPA for a student who averages the following number of hours of TV watching per day.

1. 5.0 __________ **2.** 2.1 __________

NAME ______________________ DATE __________ PERIOD __________

CHAPTER 11 Inferential Statistics

Tie It Together

Fill in the details in the graphic organizers.

Probability Distribution

mean formula: ______________

variance formula: ______________

standard devation formula: ______________

Binomial Distribution

mean formula: ______________

variance formula: ______________

standard devation formula: ______________

Normal Distribution

Emperical Rule of a Normal Distribution with mean μ and standard deviation σ.

Sketch:

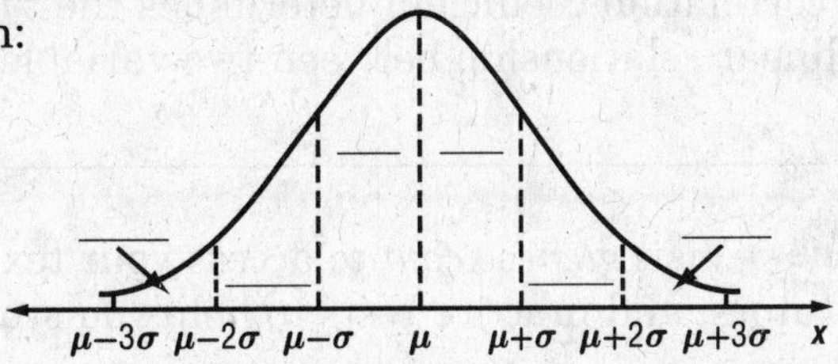

z-value for an individual data value formula: ______________

z-value for a sample mean formula: ______________

t-Distribution

t-value formula: ______________

NAME ______________________ DATE ____________ PERIOD ____________

CHAPTER 11 Inferential Statistics

Before the Test

Now that you have read and worked through the chapter, think about what you have learned and complete the table below. Compare your previous answers with these.

1. Write an **A** if you agree with the statement.

2. Write a **D** if you disagree with the statement.

Inferential Statistics	After You Read
• The median of a set of data is less affected by the presence of outliers than the mean.	
• There are three types of random variables: discrete, continuous, and infinite	
• The total area under a normal distribution curve is 10.	
• The degrees of freedom represent the number of values that are fixed after a sample statistic is determined.	
• The correlation coefficient determines the type and strength of the linear relationship between two variables.	

Math Online Visit *glencoe.com* to access your textbook, more examples, self-check quizzes, personal tutors, and practice tests to help you study for concepts in Chapter 11.

Are You Ready for the Chapter Test?

Use this checklist to help you study.

☐ I completed the Chapter 11 Study Guide and Review in the textbook.

☐ I took the Chapter 11 Practice Test in the textbook.

☐ I used the online resources for additional review options.

☐ I reviewed my homework assignments and made corrections to incorrect answers.

☐ I reviewed all vocabulary terms from the chapter.

Study Tip

- Study with a friend who is in your class. By explaining the material to someone else, you will remember it better.

NAME ______________________ DATE __________ PERIOD __________

CHAPTER 12 Limits and Derivatives

Before You Read

Before you read the chapter, think about what you know about limits and derivatives. List three things that you already know about them in the first column. Then list three things that you would like to learn about them in the second column.

K What I know...	W What I want to find out...

Note-Taking Tips

- **If you find it difficult to write and pay attention at the same time, only write down the key words that you hear.**
  Then go back later and complete your notes.
- **Review your notes as soon and as often as possible.**
  Clarify any ideas that were not complete before the chapter test.

NAME ______________________________ DATE ____________ PERIOD ____________

CHAPTER 12 Limits and Derivatives

Key Points

Scan the pages in the chapter. Write at least one specific fact concerning each lesson. For example, in the lesson on estimating limits, one fact might be that you can support your answer by using the table feature of a graphing calculator. After completing the chapter, you can use this table to review for your chapter test.

Lesson	Fact
12-1 Estimating Limits Graphically	
12-2 Evaluating Limits Algebraically	
12-3 Tangent Lines and Velocity	
12-4 Derivatives	
12-5 Area Under a Curve and Integration	
12-6 The Fundamental Theorem of Calculus	

NAME ______________________ DATE __________ PERIOD __________

12-1 Estimating Limits Graphically

What You'll Learn

Scan the lesson. Write two things that you already know about limits.

1. ______________________

____

2. ______________________

____

____

Active Vocabulary

Review Vocabulary Define *continuous function* in your own words. (Lesson 1-3)

____

____

Define *discontinuous function* in your own words. (Lesson 1-3)

____

____

Define *removable or point discontinuity* in your own words. (Lesson 1-3)

____

____

____

New Vocabulary Write the correct term next to each definition.

______________ ▶ the existence of a limit of a function $f(x)$ as x approaches c from each side

______________ ▶ the left- or right-hand behavior of $f(x)$ as x approaches c

Main Idea	Details

Estimate Limits at Fixed Values
pp. 736–740

Estimate each limit, if it exists. Then sketch the graph of each function.

Limit	Estimate	Graph
$\lim_{x \to 2} \frac{x^2 - 4}{x - 2}$		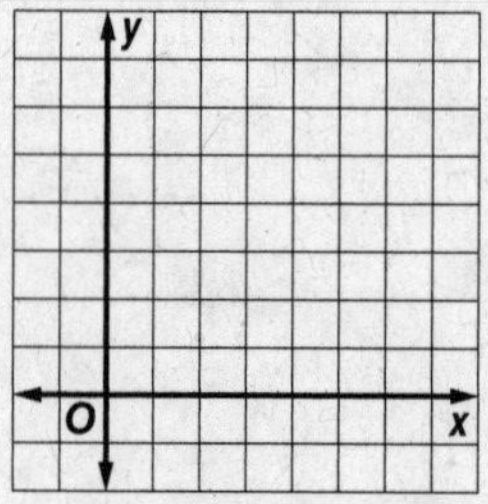
$f(x) = \begin{cases} x + 1 \text{ if } x < 2 \\ x^2 \text{ if } x \geq 2 \end{cases}$ $\lim_{x \to 2^-} f(x)$ $\lim_{x \to 2^+} f(x)$		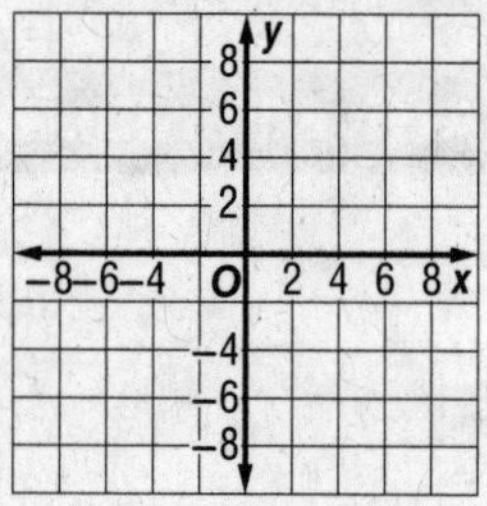
$\lim_{x \to -3} \frac{1}{(x + 3)^2}$		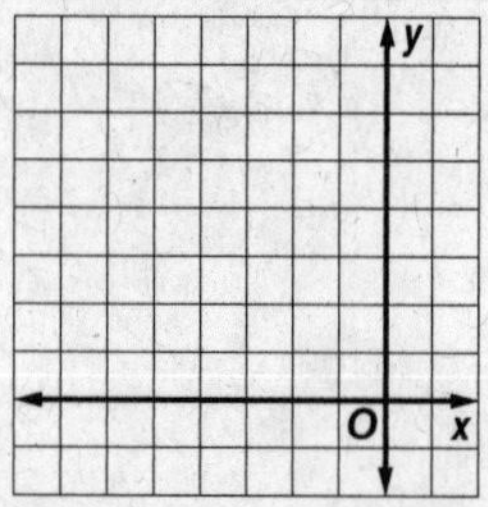

Estimate Limits at Infinity
pp. 740–742

Provide an example of a function with a limit at infinity of –3. Then sketch the graph of the function.

Function: ______________________

Graph:

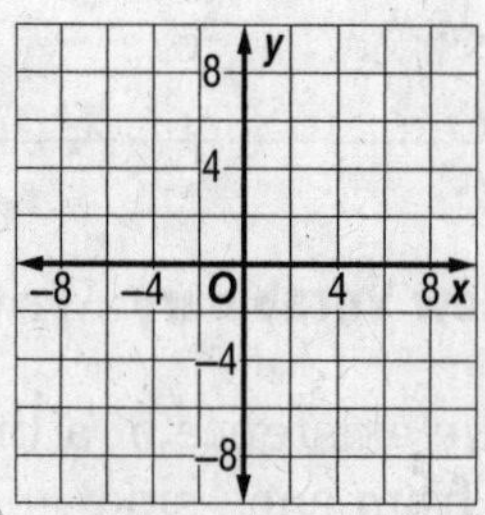

NAME ______________________ DATE ____________ PERIOD ________

12-2 Evaluating Limits Algebraically

What You'll Learn

Scan the text in Lesson 12-2. Write two facts that you learned about evaluating limits as you scanned the text.

1. ____

____

2. ____

____

Active Vocabulary

Review Vocabulary Define *limit* in your own words. (Lesson 1-3)

____

____

Define *two-sided limit* in your own words. (Lesson 12-1)

____

____

____

New Vocabulary Write the definition next to each term.

direct substitution ▶ ____

____

____

indeterminate form ▶ ____

____

____

Lesson 12-2

Main Idea	Details

Compute Limits at a Point
pp. 746–750

Evaluate each limit.

1. $\lim_{x \to 5} \frac{x^2 + 2x - 35}{x - 5} =$ ______________
 $=$ ______________
 $=$ ______________
 $=$ ______________

2. $\lim_{x \to 16} \frac{x - 16}{\sqrt{x} - 4} =$ ______________
 $=$ ______________
 $=$ ______________
 $=$ ______________
 $=$ ______________

Compute Limits at Infinity
pp. 750–753

Evaluate each limit.

1. $\lim_{x \to \infty} (3x^4 - 2x^3 + 1) =$ ______________
 $=$ ______________
 $=$ ______________

2. $\lim_{x \to \infty} \frac{x^3 - x}{2x^4 + 1} =$ ______________
 $=$ ______________
 $=$ ______________
 $=$ ______________
 $=$ ______________

NAME ______________________ DATE ____________ PERIOD ____________

12-3 Tangent Lines and Velocity

What You'll Learn

Scan Lesson 12-3. Predict two things that you expect to learn based on the headings and Key Concept boxes.

1. ______________________________

____

2. ______________________________

____

Active Vocabulary

Review Vocabulary Define *average rate of change* in your own words. (Lesson 1-4)

____

____

____

Define *secant line* in your own words. (Lesson 1-4)

____

____

New Vocabulary Match each term with its definition by drawing a line to connect the two.

difference quotient	the velocity of an object at any point in time
instantaneous rate of change	a line used to approximate the slope of a graph at a particular point
instantaneous velocity	a formula that can be used to find the slope of a secant line
tangent line	the slope of a nonlinear graph at a specific point

Lesson 12-3

Main Idea	Details

Tangent Lines
pp. 758–759

Find an equation for the slope of the graph of $y = 3x^2$ at any point.

$m =$	____________	Instantaneous Rate of Change Formula
$=$	____________	$f(x + h) = 3(x + h)^2$ and $f(x) = 3x^2$.
$=$	____________	Expand.
$=$	____________	Simplify.
$=$	____________	Factor h from the numerator.
$=$	____________	Divide by h.
$=$	____________	Sum Properties of Limits
$=$	____________	Simplify.

Instantaneous Velocity
pp. 760–761

PARKS A park ranger drops her binoculars from the top of a 400-foot observation tower. The height of the binoculars after t seconds is given by $h(t) = 400 - 16t^2$. Find the instantaneous velocity of the binoculars 4 seconds after they are dropped.

____

Helping You Remember

In your own words, describe the difference between the average velocity of a runner and the instantaneous velocity of a runner.

____

____

____

NAME ______________________________ DATE ____________ PERIOD ____________

12-4 Derivatives

What You'll Learn

Scan the examples for Lesson 12-4. Predict two things that you think you will learn about derivatives.

1. ____

____

2. ____

____

Active Vocabulary

Review Vocabulary Define *difference quotient* in your own words. (Lesson 12-3)

____

____

Define *tangent line* in your own words. (Lesson 12-3)

____

____

New Vocabulary Fill in each blank with the correct term.

derivative	The result of finding a derivative is called a(n) ______________.
differentiation	The notation $\frac{d}{dx}$ is known as a(n) ______________.
differential equation	The process of using limits to determine the slope of a line tangent to the graph of a function at any point is called the ________ of the function.
differential operator	The process of finding a derivative is called ______________.

Lesson 12-4

Main Idea	Details

Basic Rules
pp. 766–769

Find the derivative of $f(x) = -4x^3 + 2x^2 - 5$.

$f(x) =$ ______________ Original equation

$f'(x) =$ ______________ Constant, Constant Multiple of a Power, and Sum and Difference Rules

$=$ ______________ Simplify.

Product and Quotient Rules
pp. 770–771

Use the quotient rule to find the derivative of $f(x) = \dfrac{-2x^3 + 5}{x^2 + 1}$.

$f'(x) =$ ______________

$f'(x) =$ ______________

$f'(x) =$ ______________

Use the product rule to find the derivative of $f(x) = -4x^2(x^3 - 1)$.

$f'(x) =$ ______________

$f'(x) =$ ______________

$f'(x) =$ ______________

Helping You Remember

Demonstrate how to use the product rule.

____

____

____

____

NAME ______________________________ DATE ____________ PERIOD ____________

12-5 Area Under a Curve and Integration

What You'll Learn

Scan the text under the *Now* heading. List two things that you will learn in the lesson.

1. ______________________________

____

2. ______________________________

____

Active Vocabulary

Review Vocabulary Define *sigma notation* in your own words. Then identify each part of the sigma notation. (Lesson 10-1)

____

____

$$\sum_{n=1}^{k} f(n)$$

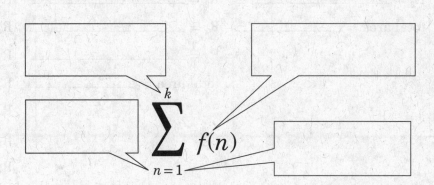

New Vocabulary Label the diagram with a term listed at the left.

definite integral

lower limit

Riemann sum

upper limit

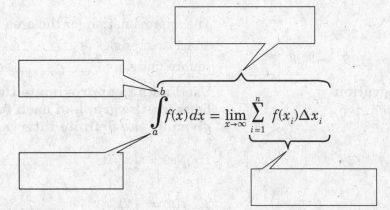

$$\int_a^b f(x)\,dx = \lim_{x \to \infty} \sum_{i=1}^{n} f(x_i)\Delta x_i$$

Lesson 12-5 *(continued)*

Main Idea	Details

Area Under a Curve
pp. 775–776

Approximate the area between the curve $f(x) = -x^2 + 18x$ and the x-axis on the interval [0, 18] using 3, 5, and 11 rectangles. Use the right endpoint of each rectangle to determine the height.

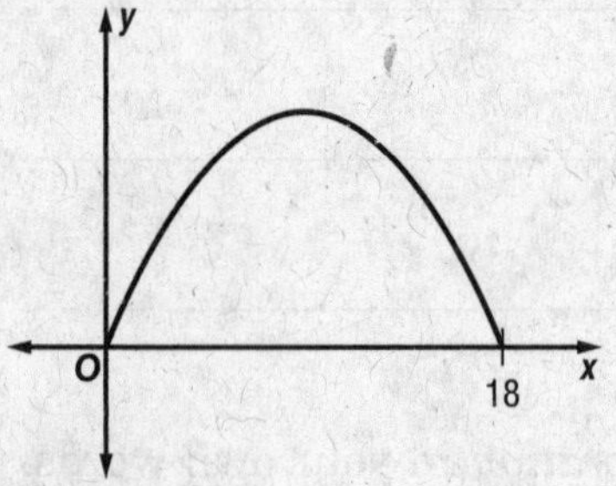

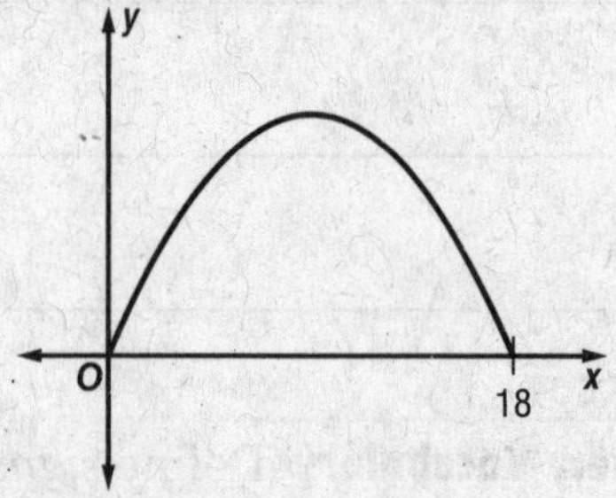

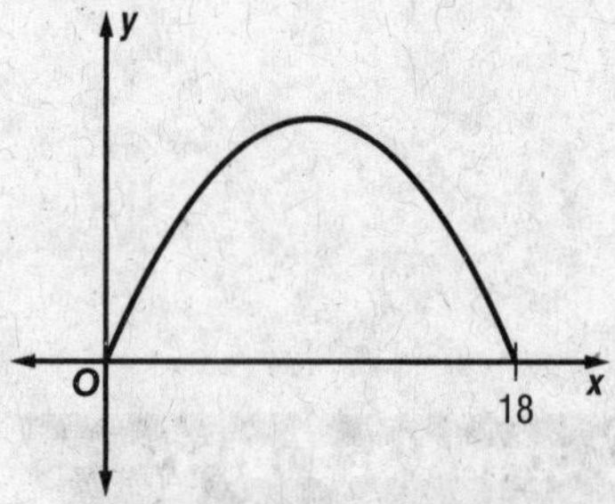

Area using 3 rectangles

R_1 = ____________

R_2 = ____________

R_3 = ____________

total area = ____________

Area using 5 rectangles

R_1 = ____________

R_2 = ____________

R_3 = ____________

R_4 = ____________

R_5 = ____________

total area = ____________

Area using 11 rectangles

R_1 = ____________

R_2 = ____________

R_3 = ____________

R_4 = ____________

R_5 = ____________

R_6 = ____________

R_7 = ____________

R_8 = ____________

R_9 = ____________

R_{10} = ____________

R_{11} = ____________

total area = ____________

The approximation for the area under the curve using 3, 5, and 11 rectangles is __________ square units, __________ square units, and __________ square units, respectively.

Integration
pp. 776–780

Use limits to approximate the area of the region between the graph of each function and the x-axis given by the definite interval.

1. $f(x) = \int_1^4 x^3\,dx$ ____________

2. $f(x) = \int_0^3 2x^3\,dx$ ____________

NAME ______________________ DATE ____________ PERIOD ____________

12-6 The Fundamental Theorem of Calculus

What You'll Learn

Scan Lesson 12-6. List two headings that you would use to make an outline of this lesson.

1. ______________________

____

2. ______________________

____

Active Vocabulary

Review Vocabulary Define *derivative* in your own words. (Lesson 12-4)

____

____

____

Define *differentiation* in your own words. (Lesson 12-4)

____

____

New Vocabulary Fill in each blank with the correct term.

antiderivatives — The ______________ of $f(x)$ is defined by $\int f(x)dx = F(x) + C$, where $F'(x) = f(x)$ and C is any constant.

Fundamental Theorem of Calculus — The function $F(x)$ is the ____________ of $f(x)$ if $F'(x) = f(x)$.

indefinite integral — The ______________________ can be used to evaluate definite integrals.

Main Idea	Details
Antiderivatives and Indefinite Integrals pp. 784–786	**Find all antiderivatives for each function.** 1. $m(x) = 3x^4$ ____________ 2. $p(x) = x^3 + 6x^2$ ____________ 3. $f(x) = \frac{1}{x^3} + \frac{6}{x^2}$ ____________ 4. $n(x) = 5\sqrt{x}$ ____________
Fundamental Theorem of Calculus pp. 786–788	**Evaluate each definite integral.** 1. $\int_1^3 (x^2 + 2x)\,dx$ ________ 2. $\int_1^9 \sqrt{x}\,dx$ ________ **Evaluate each integral.** 3. $\int (8x^2 - 2x^4)\,dx$ = ____________ = ____________ 4. $\int (x^3 - x - 1)\,dx$ = ____________ = ____________

Helping You Remember

Explain why a constant C is added to the antiderivative of a function. Provide an example of a function and two distinct antiderivatives.

____

____

____

____

NAME ______________________ DATE ____________ PERIOD ________

CHAPTER 12 Limits and Derivatives

Tie It Together

Provide each definition.

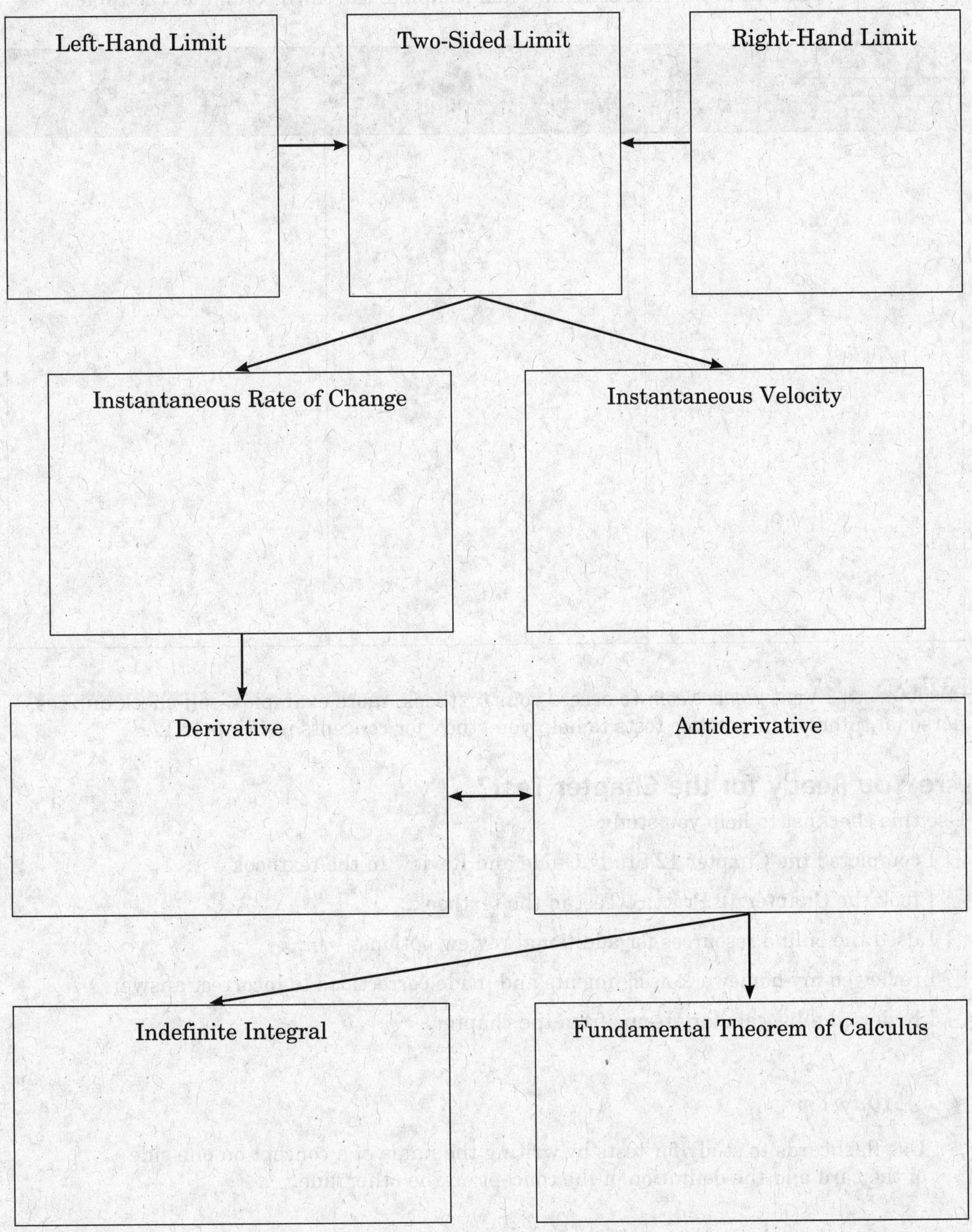

NAME ____________________ DATE ____________ PERIOD ____________

CHAPTER 12

Limits and Derivatives

Before the Test

Review the ideas that you listed in the table at the beginning of the chapter. Cross out any incorrect information in the first column. Then complete the third column of the table.

K What I know...	W What I want to find out...	L What I Learned...

Math Online Visit *glencoe.com* to access your textbook, more examples, self-check quizzes, personal tutors, and practice tests to help you study for concepts in Chapter 12.

Are You Ready for the Chapter Test?

Use this checklist to help you study.

☐ I completed the Chapter 12 Study Guide and Review in the textbook.

☐ I took the Chapter 12 Practice Test in the textbook.

☐ I used the online resources for additional review options.

☐ I reviewed my homework assignments and made corrections to incorrect answers.

☐ I reviewed all vocabulary terms from the chapter.

Study Tip

Use flashcards to study for tests by writing the name of a concept on one side of the card and the definition of the concept on the other side.